U.S. Fire Administration

Firefighter Autopsy Protocol

March 2008

 FEMA

U.S. Fire Administration

Mission Statement

As an entity of the Federal Emergency Management Agency (FEMA), the mission of the U.S. Fire Administration (USFA) is to reduce life and economic losses due to fire and related emergencies, through leadership, advocacy, coordination, and support. We serve the Nation independently, in coordination with other Federal agencies, and in partnership with fire protection and emergency service communities. With a commitment to excellence, we provide public education, training, technology, and data initiatives.

Firefighter Autopsy Protocol

■ ■ ■

Jeffrey O. Stull

International Personnel Protection, Inc.

Austin, Texas

■ ■ ■

March 2008

■ Table of Contents

■ Overview

The Firefighter Autopsy Protocol has been extensively revised since its original 1994 edition. In this new protocol, a number of additional areas of information have been provided to take into account emerging issues and new technologies as applied to the conduct of autopsies. As stated in the report, it is recommended that autopsies be performed for all firefighter fatalities where a line-of-duty death has occurred. It is further recommended that an autopsy be performed when a non-line-of-duty death may be linked to a line-of-duty exposure.

Specific sections have been provided in this report as background and areas of information as related to the conducting of autopsies. General autopsy procedures must be supplemented with additional analyses and reviews in order to ascertain specific causes and mechanisms of death and to add to the body of knowledge for understanding firefighter fatalities which, in turn, helps to prevent future firefighter fatalities. The protocol gives specific attention to several areas, where current autopsy practice may be supplemented with additional evaluations and considerations. Examples of these supplemental factors include

- ■ evaluation of victim work history with specific attention to prior exposures;
- ■ examination of personal protective equipment (PPE) for relating effects of clothing and equipment on individual parts of the body, particularly in cases of trauma and burn injury;
- ■ details in the physical examination for identifying signs of smoke asphyxiation and burn injury as contributing causes of firefighter fatality;
- ■ implementation of appropriate carbon monoxide and cyanide evaluation protocols as part of the toxicological evaluation; and
- ■ detailed toxicological evaluations where hazardous atmospheres have been encountered.

The utility of this specific firefighter autopsy protocol is intended to advance the analysis of the causes of firefighter deaths to aid in the development of improved firefighter health and safety equipment, procedures, and standards. It also is critical in helping to determine eligibility for death benefits under the Federal government's Public Safety Officer Benefits (PSOB) Program, as well as State and local programs. Lastly, implementation of this protocol will increase interest in the study of deaths as related to occupational illnesses among firefighters, both active and retired.

■ Acknowledgement

This project could not have been undertaken without the expertise of the following people who provided invaluable guidance and input throughout this project:

Professor Vincent Brannigan, JD
Dept. of Fire Protection Engineering
University of Maryland
College Park, MD 20742-3031

Dr. Merritt Birky
National Transportation Safety Board
2419 Childs Lane
Alexandria, VA 22308

Richard Duffy
Int'l Association of Fire Fighters
1750 New York Avenue, NW
Washington, DC 20006

Ms. Rita Fahy
National Fire Protection Association
1 Batterymarch Park
Quincy, MA 02169-7471

Captain Tom Hales, MD
Center for Disease Control
4676 Columbia Parkway
Cincinnati, OH 45226

Dr. James Melius
NY State Laborers
18 Corporate Woods Blvd.
Albany, NY 12211

Ms. Valerie C. Neal*
PSOB (Senior Benefits Manager)
810 7th Street, NW
Washington, DC 20531

Dr. Fredric Rieders
National Medical Services
2300 Stratford Avenue
Willow Grove, PA 19090

Kevin Roche
Phoenix Fire Department
2625 South 19th Street
Phoenix, AZ 85009

Mary Ellis**
National Fallen Firefighters Foundation
16825 South Seton Avenue, Building O
P. O. Drawer 498
Emmitsburg, MD 21727

Jim Monahan
Beebe Medical Center
424 Savannah Road
Lewes, DE 19958

*No longer with PSOB at the time of publication.
**No longer with the NFFF at the time of publication.

■ I. BACKGROUND

The U.S. Fire Administration (USFA) is committed to improving the health and safety of firefighters. This mission has created an accompanying interest in learning about the causes of firefighter deaths and injuries. In the process of researching firefighter deaths, it was determined that forensic medicine had no standard protocol that would assist a coroner or medical examiner specifically in determining the cause of a firefighter death. Many purposes for firefighter autopsy are related to ensuring benefits in addition to providing an improved understanding of fireground hazards and the effectiveness of firefighting equipment.

In 1993, the USFA initiated a project to develop a standard firefighter autopsy protocol. Experts in forensic pathology, toxicology, epidemiology, and medicolegal aspects of autopsy, as well as representatives of several national fire service organizations, were selected to serve as a Technical Advisory Committee, to provide expertise and guidance for development of the new autopsy protocol. The first firefighter autopsy protocol was finalized in 1994 and disseminated in 1995. This protocol effectively served the forensic professional and provided a basis for examining firefighter deaths more consistently.

In 2004, a revision of the firefighter autopsy protocol was undertaken to further refine and update autopsy procedures to account for new types of analyses and concerns that have arisen with respect to the conducting of firefighter autopsies. A second Technical Advisory Committee, with membership similar to the first, was formed to help guide and review the modifications. The new changes and improvements in the autopsy protocol are represented in this publication.

The consensus of the Technical Advisory Committee is reflected in the new protocol. This protocol is intended to provide guidance to medical examiners, coroners, and pathologists on uniform recommended procedures for investigating the causes and contributing factors related to firefighter deaths. The protocol recognizes and addresses those attributes of firefighter casualties which distinguish them from casualties in the general population, as well as from civilian fire casualties. These differences include the use of protective clothing and equipment, prolonged exposures to the hazardous environment, and specialized training and duties.

The accompanying documentation is intended to describe the need for a revised autopsy protocol, the situations that led to its development, and the major issues that are related to it.

I.1 Scope of the Problem

Firefighting has been described as one of the Nation's most hazardous occupations. The USFA estimated that the number of firefighters in 2005 was 1,136,650, comprising 313,300 career and 823,350 volunteer firefighters.[1] This figure included only those career firefighters working for public municipalities rather than for private fire brigades or for State or Federal government.

The NFPA defines on-duty fatalities as follows (Fahy, 2007):

On-duty fatalities include any injury sustained in the line of duty that proves fatal, any illness that was incurred as a result of actions while on duty that proves fatal, and fatal mishaps involving non-emergency occupational hazards that occur while on duty. The types of injuries included in the first category are mainly those that occur at a fire or other emergency incident scene, in training, or in crashes while responding to or returning from alarms. Illnesses (including heart attacks) are included when the exposure or onset of symptoms occurred during a specific incident or on-duty activity.

According to reports by the NFPA[2], 3,723 firefighters have lost their lives while on duty in the United States over the past 30 years (1977 through 2006); this includes the 343* firefighters who died at the World Trade Center in 2001. Excluding the World Trade Center firefighter deaths, the average number of firefighter fatalities approaches 113 per year. However, from a yearly average of 151 firefighter line-of-duty deaths in the 1970s, the average death rate has declined to 99 deaths per year since 2000. While the primary cause of line-of-duty fatalities remains sudden cardiac death, the number of such deaths per year has declined by about one-third; however, since the early 1990s the number of cardiac-related deaths has remained between 40 and 50 per year. Vehicular crashes remain the second-highest cause of line-of-duty fatalities.

Improvements in firefighter health and safety standards and practices, particularly in the areas of PPE, physical fitness, and training, are widely believed to be responsible for a significant downward trend in line-of-duty deaths during the past 30 years. Between 1977 and 2006, the Nation experienced a 43-percent drop in the annual number of firefighter line-of-duty deaths (see Figure 1). Notwithstanding the significant drop in firefighter deaths, too many firefighters die needlessly each year.

The statistical analysis of firefighter fatalities accounts for how many firefighters have died and, to some extent, explains how they died, but the available data do not explain why firefighters die. Interpreting data is made more complex by factors such as the declining number of structural fires and the year-to-year variation in number and severity of wildland fires. Moreover, a dramatic downward shift in the total number of firefighter deaths in certain years, such as 1992, 1993, and 2005, begs still more questions about what is being done correctly to prevent line-of-duty deaths.

Epidemiological studies of firefighter mortality conducted in various years provide interesting insights for comparing firefighter health and mortality rates to those of other population groups, but they, too, fall short of explaining conclusively why firefighters die, and especially why any particular firefighter dies. The interest in occupational health factors relates to the frequency of sudden deaths due to heart attacks, as well as chronic conditions which include respiratory disorders, heart disease, and cancer.

[1] USFA Web site—based on figures from the National Fire Protection Association's (NFPA) U.S. Fire Department Profile through 2005.

[2] NFPA Journal, July/Aug 2007

*The USFA shows 344 firefighters died on duty and the National Fallen Firefighters Foundation have honored 347 firefighters from the World Trade Center.

Figure 1. On-Duty Firefighter Deaths (1977-2006)[3]

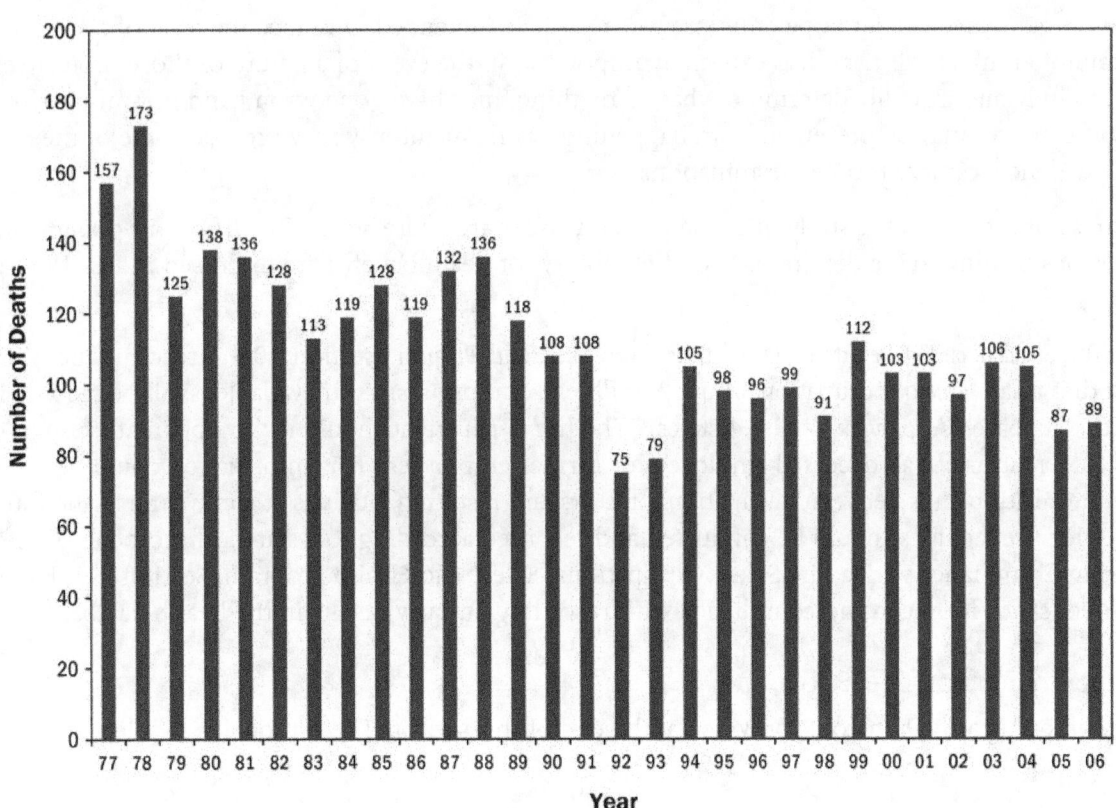

I.2 Rationale for the Protocol

The autopsy protocol was developed to give guidance to qualified professionals on the specific procedures that will be most appropriate in performing an autopsy on a deceased firefighter. The recommended procedures are intended to address the complex relationship between the firefighter and the inherently dangerous work environment where the duties of a firefighter must be performed. **It has been assumed that the user will be qualified, skilled, and experienced in performing autopsies, as the protocol is intended only to provide guidance on the special considerations that should apply to a firefighter autopsy that go beyond standard autopsy practice.**

It is anticipated that the application of this firefighter autopsy protocol will lead to a more thorough documentation of the causes of firefighter deaths and achieve three goals:

1. It will advance the analysis of the causes of firefighter deaths to aid in the development of improved firefighter health and safety equipment, procedures, and standards.

2. It will help determine eligibility for death benefits under the Federal government's PSOB, as well as State and local programs.

3. It will address an increasing interest in the study of deaths that could be related to occupational illnesses among firefighters, both active and retired.

---

[3] Fahy, Rita F., Paul R. LeBlanc, and Joseph L. Molis. *Firefighter Fatalities in the U.S.—2006.* Quincy: National Fire Protection Association, June 2007.

The work environment of the firefighter is undoubtedly one of the most inherently dangerous workplaces. To survive in that environment, the firefighter routinely uses protective clothing, respiratory protection, safety equipment, and standard operating procedures (SOPs) intended to reduce the level of risk, but which cannot eliminate all risks. It is extremely important, in the event of a failure of those protective systems, to fully and carefully determine what, if anything, may have gone wrong and how, if possible, similar occurrences may be prevented from happening again. An autopsy may provide some of the essential evidence to make those determinations.

Several areas of interest in the study of chronic health issues are addressed in Part III of this document. The specific issues relating to the determination of eligibility for death benefits are discussed in Part IV of this document.

NFPA 1500, *Standard on Fire Department Occupational Safety and Health Program*,[4] section 10.4.4 recommends, "If a member dies as a result of occupational injury or illness, autopsy results, if available, shall be recorded in the health data base." Appendix A-10.4.4 states, "The fire department should try to obtain autopsy or other medical information for all deceased employees or former employees. This information could be useful in establishing relationships between occupational factors and resulting fatalities at some time in the future. Autopsies for fire fatalities should be conducted and recorded according to a standard protocol." Annex B, Monitoring Compliance with a Fire Safety Occupational Safety and Health Program, Section B-2, Figure B-2 includes space for recording compliance with recording autopsy results in the health database.

I.3 Description of the Protocol

The Firefighter Autopsy Protocol is provided at the end of this section. The protocol is divided into the following sections:

- preliminary;
- initial examination;
- external examination;
- internal examination;
- toxicological examination;
- microscopic examination;
- summary of pathological findings; and
- conclusions.

The specific areas of procedures are described, but detailed step-by-step instructions are not provided as autopsy practice varies and changes with the specific circumstances of death.

---

[4] NFPA 1500, *Standard on Fire Department Occupational Safety and Health Program*. 2007 Ed. Quincy: National Fire Protection Association (617-770-3000; *www.nfpa.org*).

September 2007 Firefighter Autopsy Protocol

PROTOCOL	DISCUSSION

I. Preliminary
  A. Circumstances of Death
    1. Line-of-duty
      a. Fire suppression
      b. Special operations (e.g., hazmat, technical rescue)
      c. In transit to emergency
      d. Other official activity
    2. Non-line-of-duty
      a. Active firefighter, unrelated activity
      b. Former firefighter activity or exposure
  B. Medical Records Review
    1. Fire department injury/exposure records
      a. Prior incidents
      b. Prior injuries and treatments
    2. Current medical conditions/medications
      a. Prescribed
      b. Over-the-counter
      c. Administered by paramedics
  C. Complete Work History
    1. Length of fire suppression duty
    2. Other jobs held during fire service
    3. Jobs held after fire service
  D. Scene Investigation
  E. Scene Photography
    1. The body as discovered
    2. The site after the body is removed
    3. The body clothed at autopsy
    4. The body after removal of clothing
    5. Specific shots of body depending on type of injury
  F. Jurisdiction/Authority to Conduct Autopsy
	Firefighters are subject to many uncommon occupational hazards, including toxic and superheated atmospheres; explosions; falls; crushing/penetrating forces; contact with fire, electricity, or hazardous materials; and extremely strenuous and stressful physical activities.
	
	The autopsy results may be essential to determine why or how a firefighter was incapacitated, how the activity related to the cause of death, and whether protective equipment performed properly. Having a clear picture of the nature of firefighting operations that were taking place (and to which the deceased was assigned) will assist in identifying possible mechanisms of injury. If the firefighter was reported missing, try to determine the time of last contact or the length of time between the initial report and the finding of the body.
	
	The fire department should have an officer or internal Line-of-Duty Death Investigation Team assigned to conduct a death investigation. Other investigators may include the police, the State Fire Marshal (or other State officials), and/or Federal/State agencies responsible for occupational safety and health, including the National Institute for Occupational Safety and Health (NIOSH). Consult with these officials as necessary.
	
	In conducting the medical records review, obtain any documents that pertain to the incident. Document the occupational history of the deceased, including the number of years assigned as a "line" firefighter, any history of unusual exposures (or changes in frequency of exposure) to hazardous substances or diseases, and any relevant occupational medical history. Finally, all recent medical history should be reviewed, including documentation of any attempts at onscene resuscitation.

II. Initial Examination
  A. Identification of Victim
  B. Document Condition of PPE
    1. Refer to PPE diagram in Figure 8 and information in Appendix C for standardized nomenclature.
    Ppe description should include:
      a. Turnout coat
      b. Turnout pants
      c. Helmet
      d. Gloves
      e. Boots
      f. Self-contained breathing apparatus (SCBA)
      g. Personal Alert Safety System (PASS)
      h. Protective hood
      i. Clothing worn under turnouts
      j. Other PPE not listed above
    2. Use photographs to enhance documentation (see Appendix C)
	Exercise caution when handling contaminated PPE, especially from hazardous materials incidents, as residue may be harmful to those involved in the autopsy. Gloves and other PPE should be used.
	
	Ppe should be sealed in a plastic bag if fire accelerants or other volatile/toxic chemicals are suspected to be present; otherwise PPE should be air-dried and preserved for examination. (Do not store clothing wet.) Preservation of the original state of PPE, including clothing, is essential. Ppe should be considered as evidence, and handled according to the Special Incident Procedures in NFPA 1851, *Standard on Selection, Care, and Maintenance of Structural and Proximity Firefighting Protective Ensembles* (2008 edition). The Death Investigation Team should perform or assist in the evaluation/documentation of PPE condition and performance. Documentation of the chain of custody of the PPE is required, especially as it may be examined by a number of individuals; however, examinations should be limited if the condition of the clothing is fragile and will be further destroyed upon successive evaluations. Upon completion of any examination, PPE should be secured in an evidence storage area. (International Association of Fire Fighters. *Guide for Investigation of a Line-of-Duty Death.* Washington, DC, 2000).

continued on next page

PROTOCOL	DISCUSSION

C. Maintenance of Custody of Equipment
  1. Appropriate storage conditions
  2. Chain of custody
  3. Limitation of handling if clothing and equipment is fragile
	Observations and photos recorded at the scene should indicate whether the deceased was found wearing SCBA and/or other PPE. If SCBA and PASS are user-controlled, were they properly activated or working at the time of discovery of the deceased? A swab from the inside of the SCBA facepiece may help in determining operability.
	
	A qualified specialist should inspect the PPE and note any damage. NIOSH can assist in the determination of any contribution of the deceased's SCBA to the death. Ppe manufacturers may be able to assist in evaluating damage, but PPE should not be returned to the manufacturer for examination (because of concerns about product liability).
	
	Breathing apparatus filter cartridges, if any, should be retained.

III. External Examination
  A. Document Condition of Body
    1. Photograph
    2. Radiograph
    3. Record color of fingernails
    4. Record appearance of blood
  B. Document Evidence of Injury
  C. Document Evidence of Medical Treatment
  D. Collect Evidence from External Surfaces
    1. Swabs of nasal/oral soot or other substances*
    2. Hair*
    3. Injection sites
  E. Collect Vitreous Fluid
  F. Document Burns*
    1. Location
    2. Degree
    3. Etiology
    4. Percentage of body surface area (BSA)
  G. Biopsy Skin Lesions
	A complete initial examination of the body is important prior to the full autopsy, including X-rays, to help with identification, locate equipment, and look for nonobvious causes of death.
	
	Firefighters are trained to provide emergency medical care for fire casualties. Of particular importance is that resuscitative efforts for fellow firefighters are likely to be heroic and prolonged. This fact should be taken into account when examining the body for evidence of medical intervention and when interpreting the results of blood gas assay.
	
	Note the presence of soot or other unidentified substances on the skin and place samples (swabs) in a sealed container.
	
	Certain internal samples (such as soot swabs and vitreous fluid) which can be done before the body is opened are taken at this point because collection can be accomplished in a more controlled manner, thus reducing the potential for cross-contamination of the surfaces.
	
	Hair samples should be about the thickness of a finger, pulled out so as to include the roots, tied around the middle, with the proximal and distal ends marked, and stored in a plastic evidence bag. Match burn injury locations to areas of heat/thermal damage on clothing and equipment.
	
	Vitreous fluid should be taken from both eyes. Vitreous fluid can be used to corroborate blood alcohol levels.

IV. Internal Examination
  A. Document Evidence of Injury
  B. Document Evidence of Medical Treatment
  C. Describe Internal Organ System
  D. Collect Samples for Toxicologic Analysis
    1. Blood (2 x 20 cc red- and grey-top tubes)
    2. Urine (20 to 30 cc) and/or trimmed bladder
    3. Bile (all available) or gallbladder (if bile unavailable)
	Soot swabs should be obtained from the upper and lower airways as well as from the inside of the SCBA facepiece. These will assist in the determination of SCBA usage and operability. Note any unusual odors/colors of anything found during the internal examination. Fresh-frozen samples of vital organs should be taken and retained a minimum of 90 days, preferably longer, as storage space permits.
	
	An area of growing interest is the cancer rate of firefighters. Potentially cancerous tissue should be biopsied and saved. Additionally, histological type and the exact location of the tumor (if site-specific) within an organ should be documented in detail.

*May not be required for clear traumatic death

continued on next page

PROTOCOL	DISCUSSION

4. Cerebrospinal fluid (up to approx. 30 ml)
5. Soot swabs from airway*
   a. Tracheal
   b. Bronchial
6. Representative sampling of gastric and duodenal contents (50 g; note total amount)
7. Take and retain fresh-frozen samples
   a. Lung 100 g
   b. Kidney 100 g
   c. Liver 100 g
   d. Spleen 100 g
   e. Skeletal muscle (psoas or thigh) 20 g
   f. Subcutaneous fat 20 g
   g. Section of bone with marrow (3-4 cm)
   h. Brain 100 g
8. Additional specific samples to be taken:
   a. Tied-off lower lobe of right lung (store in arson debris paint can)
   b. Peripheral blood from leg vein (fluoridated and red-top tubes)
   c. Any specimens taken in field or during hospital resuscitation
   d. Sample hematomas
   e. Any other sites should be labeled

In the case of incinerated remains, bone marrow or spleen may be the only source of tissue for toxicological studies, especially for those establishing carbon monoxide levels. Request determination of carbon monoxide content and of carbon monoxide-binding capacity of mixture from water extract of spleen, kidneys, or other organs. Gastric and duodenal contents should be representative. Solid dosage forms should be removed, counted, and analyzed.

When taking lung samples, use the right lung because aspirated foreign materials have a greater propensity to lodge in the right lung. Soot particles and other heat injuries indicate that the victim was breathing in fire. Absence of soot particles does not prove that the victim was already dead when exposed to the fire.

V. Toxicological Examination
A. Urine Screen/Analysis
1. Volatile compounds (e.g., benzene, hydrocarbons including accelerants, ethanol)
2. Psychoactive substances (e.g. opiate derivatives, marijuana metabolites, cocaine metabolites, stimulants, phencyclidine)
B. Blood Analysis.
1. Carboxyhemoglobin, methemoglobin, sulfhemoglobin
2. Volatile compounds (see A.I. above)
3. Other (e.g., hydrocyanic acid, flouride)
4. Confirm results of positive urine screen
C. Subcutaneous Fat Analysis
1. Organic compounds, including:
   a. Herbicides
   b. Pesticides
2. Polychlorinated biphenyls (PCBs)
D. Soot Screen (from swabs)*
1. Metals, including:
   a. Arsenic
   b. Antimony
   c. Lead

The toxicologic analysis performed for firefighters should be of a higher order than that performed for civilian fire casualties. In addition to ascertaining blood levels of various toxic products that are commonly found in a fire environment, it is beneficial to know about the presence of any judgment-impairing substances. This may be important in the determination of eligibility for death benefits as well as for determining causality.

Determinations of asphyxiation from carbon monoxide levels should take into account victim medical history (i.e., smoking) in addition to other types of exposure. If victim survived carbon monoxide poisoning for several hours, portmortem samples usually will fail to show presence of carboxyhemoglobin. Blood taken at time of admission to hospital still may be available and of particular value.

Determination of specific levels of metals, organic compounds, and gross particulate matter should be conducted because firefighter exposure to these substances is believed to be greater than that for civilians. Additionally, this information may yield important clues about the cause, manner, and mechanism of firefighter death.

Use vitreous fluids or bile to confirm presence of ethanol in either blood or urine.

* May not be required for clear traumatic death

continued on next page

PROTOCOL	DISCUSSION

2. Organics, including:
   a. Pesticides
   b. Herbicides
   c. Vinyl chloride
   d. Acrylonitrile
   e. Acrolein
3. Particulate analysis (e.g., asbestos)
	Use caution when noting the presence of hydrocyanic acid, as it can be produced by bacterial decomposition within the tissues of the deceased.
	
	Check for the presence of PCBs and polynuclear aromatic compounds in the subcutaneous fat, as this will help in the determination of a history of exposure.

VI. Microscopic Examination A. Findings of Microscopic Examination	Representative samples of all organs and body systems should be collected. The sections should be microscopically examined for malignant neoplasms and other abnormalities, including suggestive premalignant changes

VII. Summary of Pathological Findings
   A. Medical Facts
      1. Correlation
	State objective findings related to gross and microscopic examinations. Correlate physical circumstances, toxicological analyses, and other investigative studies to pathological findings.

VIII. Conclusions
   A. Discrepancies
      1. Inconsistent observations
      2. Differences between death certificate and subsequent findings
   B. Conclusions
      1. List diagnoses on a separate page
      2. Cause and manner of death
	Include determination of **cause, manner, and mechanism** of death. Describe discrepancies between evidence collected or observations of eyewitnesses and the autopsy findings.

II. Medicolegal Autopsy Procedures in the United States

The need to investigate and understand the cause of death, particularly when it occurs under unusual, confusing, or ambiguous circumstances, is almost universal. Nearly every country has established requirements for the medicolegal investigation of unforeseen, unnatural, or violent deaths, usually including workplace accidents and job-related deaths. However, unlike some other industrialized nations, no national system of death investigation exists in the United States. Death investigation in the United States falls under the authority of State and local officials

Legal structures governing death investigation vary considerably among the 50 States, the District of Columbia, and the territories. Depending on the jurisdiction, the official responsible for determining the cause, manner, and mechanism of death may be a coroner or medical examiner. Eleven States operate coroner systems (either district or county coroners). Eighteen States use a State, district, or county medical examiner system. Eighteen States operate under a mixed system of State or county medical examiners and county coroners/medical examiners.[5] Appendix A lists the specific practices used in each State.

Most firefighter deaths are investigated as unusual or unforeseen deaths according to State laws and regulations, and a high level of discretion is afforded to coroners and medical examiners in the manner of fulfilling their duties and responsibilities. Only one State, Maryland, specifically requires a medicolegal investigation of all firefighter deaths and, in fact, has a staff epidemiologist to study firefighter deaths. Other States such as New Jersey have designated the Division of Fire Safety as the lead agency for investigating fire service accidents, but have established no autopsy requirements.

Three publications attempt to organize and describe medicolegal autopsy requirements in the United States:

1. Wecht, C.H. United States Medicolegal Autopsy Laws. 3rd ed. Arlington: Information Resources Press, 1989.
2. Combs, D.L., R.G. Parrish, R. Ing, et al. Death Investigation in the United States and Canada, 1995. Atlanta: Centers for Disease Control, U.S. Health Services, 1995.
3. Ludwig, Jurgen. Autopsy Practice. 3rd ed. Totowa: Humana Press, 2002.

Notwithstanding the differences among the various systems, all death investigation systems are intended to respond to questions of who died, how and why a death occurred, and (as applicable) who is responsible for the occurrence. This information, in turn, may be used in legal proceedings; to compile vital statistics; to evaluate medical care and treatment; and to compile factual information on clinical, anatomical, pathological, physiological, and epidemiological subjects for research purposes.

---

[5] www.cdc.gov/epo/dphsi/mecisp/death_investigation_in_the_united_states_and_canada.htm

II.1 When Is an Autopsy Required?

An autopsy is not performed as a part of every death investigation. In most cases, the determination of the need to perform an autopsy is a discretionary responsibility of the coroner or medical examiner. The issuance of a death certificate does not require an autopsy, and only a death certificate is needed to qualify for most insurance and death benefit programs. The coroner or medical examiner may determine that no autopsy is required in any situation where there is sufficient other evidence to make conclusive determinations on the cause and manner of death. In past years, an autopsy typically was omitted when the firefighter death was believed to have been caused by natural causes, such as cardiac ischemia, even when it occurred on the scene of, or en route to or from, a fire or emergency incident (see Goode, 1990). However, autopsies are now recommended for all firefighter deaths, and the International Association of Fire Fighters (IAFF) and other organizations encourage this practice. NIOSH's Fire Fighter Fatality Investigation and Prevention Program (FFFIPP) uses autopsy results in the analysis of firefighter fatalities and specifically recommends that an autopsy be performed even if the cause of death is presumed to be natural. Section III.3, Investigation of Line-of-Duty Deaths, provides additional details on this NIOSH program.

Many coroners and medical examiners have had to limit the number of autopsies performed because of cost and time constraints. Fiscal pressures have increased as the number of death investigation cases has increased, particularly those involving violent deaths. The cases in which an autopsy is most likely to be omitted include those where there is a known and undisputed cause of death without suspicion of criminal activity; line-of-duty deaths often fall within these parameters. Autopsies are sometimes omitted because of the religious or personal preferences of the deceased and his or her family.

The failure to conduct autopsies appears to be of significant concern throughout the medicolegal community. Performing autopsies, even in cases of prolonged illness or involving individuals with prior medical histories, would be valuable in conclusively determining the cause of death, gaining a more detailed understanding of injury and disease processes, and evaluating the quality of medical care. According to some in the death investigation profession, a decline in the level of interest in pathology and forensic pathology among medical students has led to a shortage of trained professionals to conduct these procedures.

Autopsies usually are performed to establish or verify the cause of death, or to gather information or evidence that would be helpful in an investigation. Without an autopsy, specific causes, contributing factors, and underlying conditions may go undiscovered and unreported. In the case of firefighter fatalities, this lack of information may hamper industry and fire service understanding of the hazards of firefighting significantly, and limit the ability to develop more effective ways to prevent firefighter deaths and injuries.

II.2 Definition of Manner, Cause, and Mechanism of Death

Because the firefighter death certificate and autopsy results have legal ramifications, it is important to clarify the differences among manner, cause, and mechanism of death, and especially to recognize that various entities may use the terms "cause of death" or "nature of death" in ways quite different from their appropriate use in the medicolegal autopsy.

■ **Manner of death** refers to classification of the death as natural, accidental, homicidal, or suicidal.[6]

The phrases "cause of death" and "mechanism of death" often are used interchangeably by clinicians and laymen, but they are not synonymous:

---

[6] Ludwig, Jurgen. *Autopsy Practice.* 3rd ed. Totowa: Humana Press, 2002.

■ **Cause of death** is the "disease or injury that sets in motion the physiologic train of events culminating in cerebral and cardiac electrical silence."[7]

■ **Mechanism of death** is the "physiological derangement set in motion by the causes of death that leads to cessation of life."

Thus, for example, a firefighter who dies from a cardiac event at the scene of a fire may have "athero-sclerotic heart disease" as the "cause of death" and "ventricular arrhythmia" as the "mechanism of death."

II.3 Chain of Custody and Documentation

Careful documentation is essential both because of the legal ramifications, and the medical and epidemiological issues surrounding firefighter deaths. Documentation to maintain the chain of custody is of particular importance in medicolegal cases. The following recommendations are made for the information that is included with specimens that are submitted for toxicological studies:

■ information that identifies each specimen, including the site where taken;

■ specific details about the requested analytical test methods;

■ relevant information about circumstances surrounding the specimen (e.g., emergency room measures that could affect certain drug levels); and

■ signatures that document the chain of custody.

Shipping of autopsy specimens requires special consideration. Specimens must be packaged appropriately to guard against breakage and to ensure the integrity of the samples. Tissues or body fluids submitted for analysis for volatile substances should be packaged in glass rather than plastic, although plastic may be acceptable for other samples. Any caps, lids, stoppers, or other loose parts of a container should be taped into place. The materials used in shipping the specimens must not compromise the samples (e.g., paraffin blocks should be not be wrapped in cotton because cotton fibers could adhere to the paraffin). Containers with wet or frozen samples must be packaged inside a second container that includes material sufficient to absorb all liquid in case of leakage. Frozen samples require ice or dry ice around the sample and in the secondary container; an insulated mailing container is necessary.

The *Autopsy Handbook*[8] specifies that "medicolegal material is sent by messenger, registered mail, or air express" and that specimen labels (inside the shipping container) include the following information:

■ name and address of sender;

■ name and address of recipient;

■ description of container and source and nature of contents;

■ tag stating that the shipment is evidence; and

■ detailed requests for specific examinations.

Containers should be sealed with a tamper-proof method, such as with sealing wax imprinted with the sender's thumbprint. The outer mailing container should have address labeling and also appropriate labels such as:

---

[7] Ibid.

[8] Ludwig, op.cit.

■ Biohazard;

■ Perishable Material;

■ Fragile, Rush, Specimen; and/or

■ Glass, Handle with Care.

Postal regulations must be followed. The publication "Domestic Mail Manual" is updated by the Postal Service periodically. It is recommended that the receiving party be notified by telephone or electronic mail when a shipment is initiated.

II.4 Retention of Autopsy Specimens and Paperwork

Autopsy specimens should certainly be retained until the investigations into the firefighter's death have been completed and any litigation surrounding the firefighter's death has been resolved. After that point, specimens and paperwork should be retained for a significant period of time. Recommended minimum storage times, **following the completion of the investigation**, are listed in Table 1.

Table 1. Recommended Minimal Storage Times for Autopsy Specimens and Paperwork[9]

Sample Type	Minimum Retention Time
Wet tissues	6 months
Accession records	1 year
Quality assurance documents	2 years
Paraffin blocks and photographs	5 years
Autopsy authorization forms	7 years
Autopsy reports and slides	20 years

It is important to recognize that each State or local medical examiner or coroner's office may have retention times that are very different from the recommendations provided above. In fact, these offices may have procedures in place that require automatic disposal of certain records or samples that will require extraordinary efforts on the part of fire departments or other individuals for continued storage and maintenance. Of principal concern is the retention of samples until after any investigation is completed. Further, certain statutes of limitation for potential litigation are likely to extend beyond the investigation period. In these instances, it is important to file a request for extending the retention times for specific samples with the State or local medical examiner or coroner. In certain circumstances, it may be necessary to identify alternative storage locations that meet all the storage requirements for autopsy samples complete with detailed chain of custody.

---

[9] Ibid.

■ III. Occupational Aspects of Firefighting of Specific Concern to Autopsy

Firefighter fatalities often result from complicated scenarios. Due to the nature of the occupation, a firefighter's death could be caused by a wide variety of single factors or a combination of several factors. For example, a firefighter could die from a stress-induced heart attack caused by simple over-exertion; or a firefighter could die from asphyxiation which is actually caused by the failure of his or her breathing apparatus; or a firefighter could die from hypothermia, resulting from being trapped in a structural collapse while fighting a fire on an extremely cold day. A firefighter's death could be caused by the inhalation of toxic products of combustion, burns, traumatic injury, exposure to hazardous materials, radiation, a variety of other singular causes, or a combination of factors.

A better understanding of the actual causes of firefighter deaths, including all of the causal factors, will require a thorough examination of the protective clothing and equipment that are involved in the incident, a detailed analysis of the situation, and the details, such as carboxy-hemoglobin levels and the presence of toxic products in the respiratory and circulatory systems, that can be obtained only through an autopsy.

III.1 Firefighter Death Classification

The USFA and the NFPA separately publish annual analyses of firefighter line-of-duty deaths. Although the two entities share information, they collect data independently. The NFPA identifies cases of firefighter fatalities through sources including newspaper reports and Internet sites such as firehouse.com. The NFPA then makes contact with the department (after the funeral) and collects information about the fatality.

To describe the mechanism of injury, the NFPA uses categories based on coding used in the 1981 edition of NFPA 901, *Uniform Coding for Fire Protection*. The NFPA studies incident reports and witness accounts as available, and then determines which classification best describes that individual fatality. The nine causal categories on which the system is based include

1. Fell/Slipped.
2. Struck by.
3. Overexertion/Strain.
4. Fire Department Apparatus Accident.
5. Caught/Trapped.
6. Contact with/Exposure to.
7. Exiting or Escaping/Jumped.
8. Assaulted.
9. Other.

In any particular year, the categories used in summary reports do not include all of the above categories. The NFPA analyst makes classifications based on the actual events for that year, and the reported causal categories may vary from year to year. In certain years, some categories which had extremely few events may be grouped into the "other" category.

While cardiac arrest and other stress-related fatalities are the leading cause of fireground deaths, this classification system does not differentiate the causes of cardiac and stress-related cases; all are classified as "Overexertion/Strain." Although firefighting is widely recognized as a highly stressful occupation, the physiological and psychological effects of job-related stress have not been clearly established or differentiated, particularly as they affect mortality and morbidity.

The annual reports also describe firefighter fatalities according to the nature of the death (i.e., the medical cause of death), using the following categories:

- Sudden cardiac event
- Internal trauma
- Asphyxiation
- Crushing
- Burns

- Drowning
- Stroke
- Electrocution
- Hemorrhage
- Gunshot

- Aneurysm
- Fracture
- Heat stroke
- Pneumonia
- Other

Depending on the fatalities sustained for that year, the categories included in the report may not include all of those listed above. In addition, new categories may be created to reflect different circumstances.

Because the reported categories may vary from year to year, one must be careful when comparing results from year to year. For example, if there are several drowning deaths in one year, those would likely be reported as a separate category in the annual report;however, if there were only a single drowning in the next year, then likely that death would be included as part of the "other" category in that year's report. Therefore, upon request, NFPA is willing to analyze data for particular situations.

It should be noted that these categories do not correspond with International Classification of Disease (ICD-10, released July 2007) or SNOMED (Standardized Nomenclature of Medicine) cause categories. There are also new classifications of death and injury as the result of terrorism incidents that have been established by the National Center for Health Statistics of the Department of Health and Human Services.

III.2 Trends in Line-of-Duty Deaths

The overall downward trend in line-of-duty deaths has been driven primarily by the downward trends in deaths attributed to cardiac arrest and in deaths during fireground operations or while at the fire scene. Fireground deaths account for more than half of all firefighter duty deaths over the last 30 years; in 42.9 percent of the cardiac-related deaths, the firefighters' cardiac symptoms appeared during fireground operations. The downward trend in the number of fireground deaths has corresponded with a downward trend in the number of structural fires, although in recent years the death rate has continued to decline while the number of structural fires has held steady (see Figure 2). Death rates due to traumatic injuries (smoke inhalation, burns, and crushing or internal trauma) injuries remain a significant concern.

Other areas of concern for the period 1977 through 2006 include the following[10]:

- Wildland firefighting accounted for 338 fatalities, and aircraft crashes contribute significantly to this number.
- Road vehicle crashes accounted for 406 fatalities, mostly volunteer firefighters, and are the second greatest cause of firefighter fatalities.

---

[10] Fahy, Rita, Paul LaBlanc, and Joseph Molis. "Firefighter Fatality Studies 1977-2006." NFPA *Journal*, July/Aug. 2007.

- Falls from apparatus while en route to or from alarms accounted for 41 deaths in the first 10 years, but only 4 in the years 1999-2006 (and none in most of the 1990s).
- Training deaths accounted for 247 fatalities, and the number of deaths in the most recent decade is nearly twice that of the first decade in this time period.

Figure 2. Trend for Firefighter Deaths at Structure Fires (courtesy of NFPA)

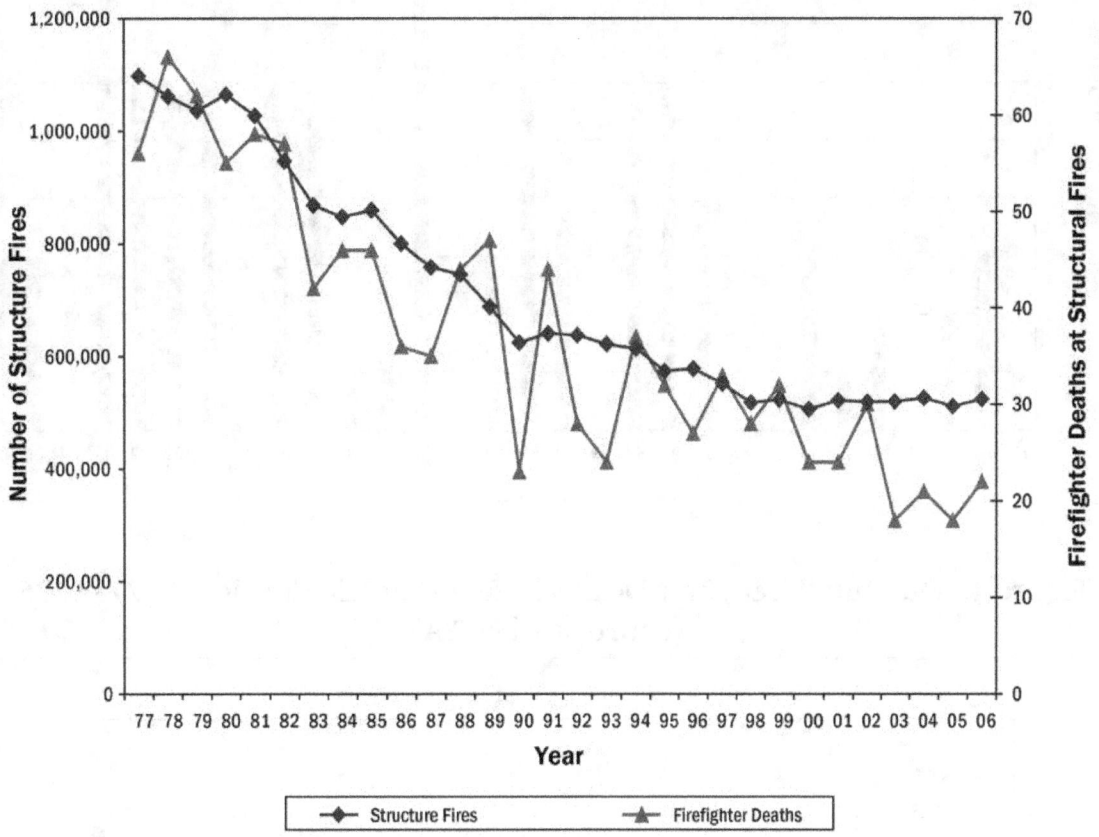

Figures 3 to 5 show these trends for the period 1977 to 2006.

Sudden cardiac death leads all categories of line-of-duty deaths. Between 1977 and 1991, 45 percent of all firefighter deaths resulted from cardiac disorders, most from myocardial infarction. The proportion of deaths resulting from heart attacks has varied from 33.6 percent to 53.9 percent over the 15-year period. Fahy (2007) reports that the number of deaths has remained between 40 and 50 per year for the period since the early 1990s, although the year 2006 saw a record low of only 34 sudden cardiac deaths.

continued on next page

**Figure 3. Firefighter Deaths in Aircraft Crashes Related to Wildland Fires (1977-2006)
(courtesy of NFPA)**

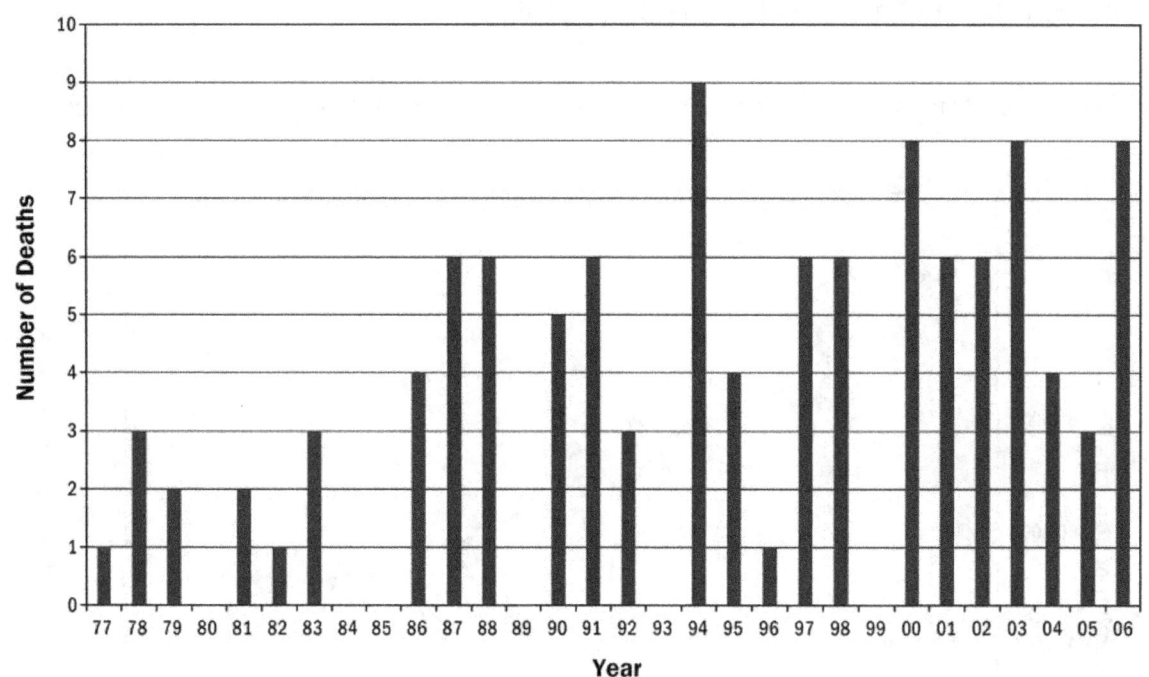

**Figure 4. On-Duty Firefighter Deaths in Road Vehicle Crashes (1977-2006)
(courtesy of NFPA)**

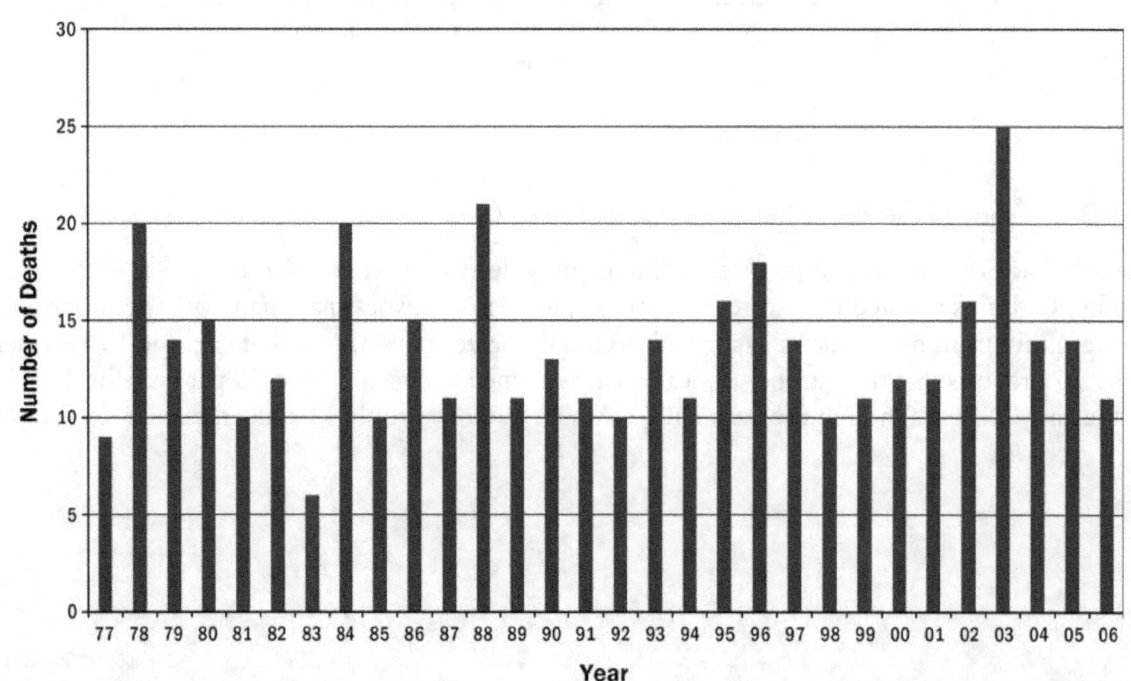

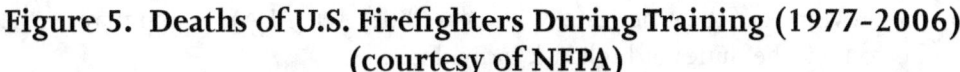

Figure 5. Deaths of U.S. Firefighters During Training (1977-2006) (courtesy of NFPA)

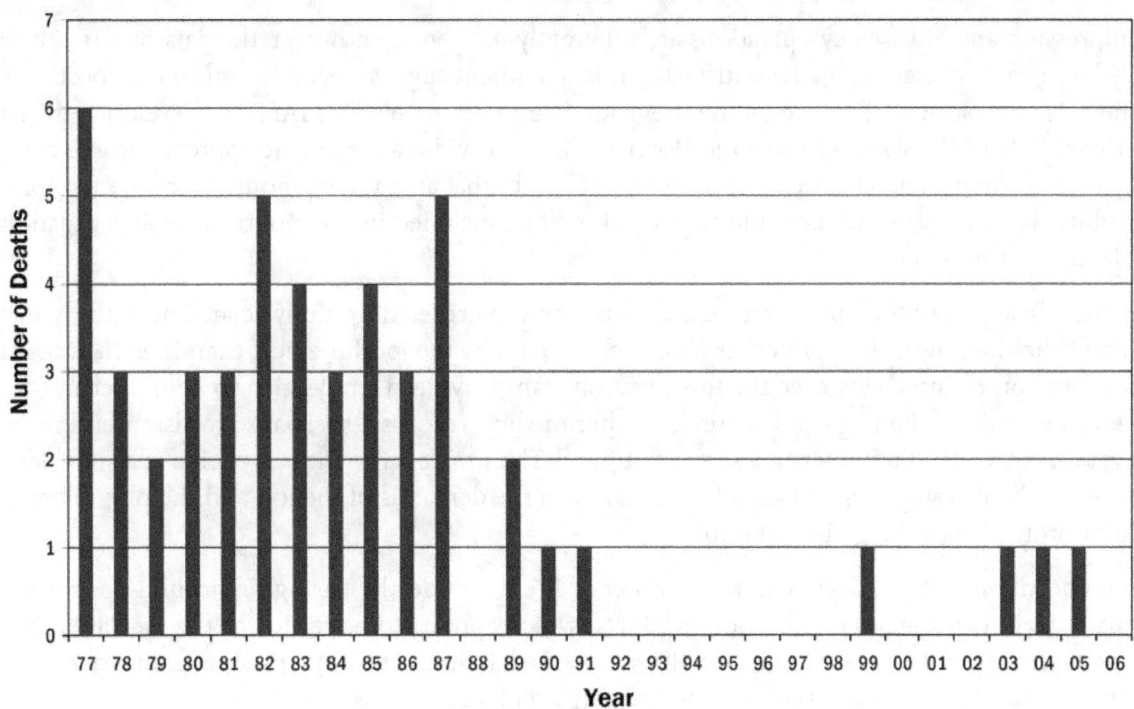

That same report points out that, according to NIOSH, "Firefighting activities are strenuous and often require firefighters to work at near maximal heart rates for long periods. The increase in heart rate has been shown to begin with responding to the initial alarm and to persist throughout the course of fire suppression activities." Fahy's report also refers to a study published this year by Kales, et al.[11]; in which the risk of dying during specific fire department duties was evaluated. The risk of death due to coronary heart disease was 10 to 100 times higher during firefighting activities than during nonemergency duties.

Fahy (1993) reported that an NFPA study of fatal firefighter heart attacks conducted for the USFA determined that about 40 percent of the firefighters who died on duty from heart attacks between 1981 and 1990 (and for whom medical documentation was available) had prior histories of cardiac ischemia, myocardial infarction, or coronary artery bypass surgery. An additional 39 percent had prior histories of acute atherosclerosis (defined as more than 50 percent occlusion); most of these cases involved occlusions greater than 70 percent. Any of these conditions could have represented sufficient cause for disqualification from continued firefighting duty under the provisions of NFPA 1582, *Standard on Comprehensive Occupational Medical Program for Fire Departments*, which was adopted in 1992.

The adoption of health maintenance and physical fitness requirements for firefighters is a controversial subject and the requirements of NFPA 1582 have not been widely adopted. This subject is further

---

[11] Kales, Stephanos, Elpidoforos Soteriades, Costas Christophi, and David Christiani. "Emergency Duties and Deaths from Heart Disease among Firefighters in the United States." *New England Journal of Medicine*, Vol. 356 (12), 2007, pp. 1207-1215.

complicated by the provisions of the Americans with Disabilities Act (ADA), which may restrict the ability of fire departments to limit the duties of high risk individuals.

III.3 Investigation of Line-of-Duty Deaths

Fire suppression and emergency operations are inherently dangerous; however, the data on firefighter line-of-duty deaths presented by the statistics in this document suggest that a significant proportion of firefighter deaths, particularly those on the fireground, are preventable. The IAFF has developed the *Fire Fighter Line-of-Duty Death and Injury Investigations Manual,*[12] which provides a systematic approach to the overall investigation of fireground fatalities. The IAFF Manual notes that an autopsy should be requested for every line-of-duty death and the results of the autopsy should be included in the report of the investigation. The IAFF Manual further states

The medical histories of firefighters are best analyzed and interpreted by a physician. The autopsy of a firefighter who died in the line of duty is always conducted by the local medical examiner, in accordance with accepted procedures. However, the investigation team may need a physician to help interpret the medical examiner's findings and/or review other medical records. Physicians are also useful in investigations in which firefighters are severely injured. The fire department's physician is a suitable candidate for the investigation and already familiar with the demands of the job and the physiological burden of protective clothing and equipment.

There has been a significant decline in the number of firefighter deaths during fireground operations, particularly from exposure to combustion products, which appears to be related to the increased use of better protective equipment. Firefighter deaths due to cardiac ailments remain a significant concern, as do traumatic injuries from vehicle accidents and training accidents.

Evaluating the thermal performance of various types of firefighter protective clothing is an example of an area where considerable insight can be gained through accurate anatomical descriptions obtained from an autopsy. Toxicological studies can help investigators better understand the effectiveness of SCBA use and operating procedures on preventing fireground exposures to hazardous atmospheres. Evaluations of body fat, muscle development, and special coronary studies can help develop a database on the relative fitness of firefighters. These types of studies will help reinforce lessons that should help the fire service improve fireground operating procedures, protective equipment, training, and physical fitness. They also can help support the development and use of criteria for regular medical evaluations for firefighters.

If the number of line-of-duty deaths continues to decline, it will become more difficult to evaluate improvements in firefighter safety through the mortality statistics. This will place increased emphasis on the need for a detailed investigation and documentation of each and every line-of-duty death. It is a matter of compelling public interest that information about the cause and manner of all firefighter line-of-duty deaths should be thoroughly and systematically collected. The autopsy results should be an important part of the record in each case.

In 1998, the National Institute for Occupational Safety and Health (NIOSH) instituted the Fire Fighter Fatality and Prevention Program (FFFIPP). The FFFIPP investigates firefighter line-of-duty deaths (and selected non-fatal injuries) with the goal of formulating recommendations for the prevention of future casualties. The investigators use the Fatality Assessment and Control Evaluation (FACE) model.

---

[12] *Fire Fighter Line-of-Duty Death and Injury Investigations Manual* (updated in 2000). International Association of Fire Fighters (IAFF), 1750 New York Avenue, NW, Washington DC 2006. (202-737-8484; *www.iaff.org*).

Medical records, death certificates, and autopsy reports as well as interviews and evaluations of personal protective equipment (PPE), particularly SCBA, are integral to the process of investigating fatalities. Each report includes a summary of the incident and specific recommendations for preventing similar events. Reports omit department and individual identifiers, as the focus in not on determining fault or blame, but rather on understanding the causes of firefighter fatalities and then developing and disseminating recommendations for prevention of fatalities. An examination of each NIOSH firefighter fatality report was conducted with the specific focus of discerning specific comments related to the conduct of autopsies. These results are provided in Appendix C.

III.4 Alcohol and Drugs

A relatively routine examination as part of any autopsy is an alcohol and drug screen. These analyses are provided as part of toxicology reports. Toxicology reports in most autopsies document the positive and negative findings of a series of tests conducted to detect specific substances that may have caused death. Such tests commonly include tests for the presence of pharmacological agents and illegal drugs. Blood tests for the presence of ethyl alcohol are conducted to determine whether the deceased was under the influence of an intoxicating beverage at the time of death. In the case of fire victims, the toxicology report should include analyses of blood, urine, other body fluids, and tissues for the presence of combustion products and other toxicants and their biomarkers (see section below), as well as alcohol and drugs.

It is extremely important that, in the determination of alcohol levels, the effects of postmortem changes and specimen storage be accounted for. Blood alcohol concentrations obtained at autopsy are valid until putrefaction begins. This may vary from several hours to a few days, depending on the environment. Most autopsy procedures recommend the addition of sodium fluoride at a concentration of 10 mg/mL of blood to the sample and the storage of the sample in a refrigerator. Considerations for evaluation of blood alcohol levels include

- If the blood is analyzed soon after withdrawal or if the blood is kept in the refrigerator, results usually are reliable, even if no sodium fluoride has been added.

- If the air space about the blood sample is large, alcohol can evaporate and a falsely low blood alcohol level can result.

- Putrefaction changes before autopsy or during storage may cause a falsely high blood alcohol concentration. Ethanol can be produced in the specimen container, usually in the absence of a preservative, as the fluoride inhibits bacteria far more effectively than fungi. Higher fluoride concentrations are required for inhibiting fungal growth.[13]

- Although there is no major difference in alcohol concentrations of blood samples from the intact heart chambers and the femoral vessels, autopsy samples from pooled blood in the pericardial sac or pleural cavity are unsatisfactory; blood should be withdrawn from peripheral vessels.

- Blood alcohol concentrations vary from vitreous, urine, or tissue samples as compared to alcohol determined through stomach contents. These variations depend on whether blood alcohol concentrations were increasing or decreasing at the time of death.[14]

---

[13] Harper D.R. and J.E.L. Correy. "Collection and storage of specimens for alcohol analysis." In *Medicolegal Aspects of Alcohol*, J.C. Garriott, ed. Phoenix: Lawyers and Judges Publishing Co., 1997, pp. 145-169.

[14] Caplan Y.H., "Blood, urine and other tissue specimens for alcohol analysis." In *Medicolegal Aspects of Alcohol*, J.C. Garriott, ed. Phoenix: Lawyers and Judges Publishing Co., 1997, pp. 74-86.

The principal drugs for analysis include those for common narcotics, barbiturates, amphetamines, hallucinogens, or cannabinoids. Tests for other prescription and nonprescription drugs are performed occasionally to detect such compounds as common steroids, analgesics, and other indicators of coexisting illnesses/conditions, as well as of drugs used in emergency resuscitation attempts. Methods typically used in these analyses are described in Table 2.

Table 2. Common Methodologies for Toxicological Analysis

Technique	How Used
Volatiles by Gas Chromatography (GC)	Usually used for testing ethanol content; testing is applied as part of a general panel to detect and quantify numerous volatile compounds that include methyl, ethyl, and isopropyl alcohols and ketones. t-Butyl alcohol is used as an internal standard because it does not occur naturally.
Specific drug screening by Enzyme-Multiplied Immunoassay (EMIT)	EMIT can detect but not quantify dependence drugs of abuse. Specific test panels are available for cocaine metabolites, tricyclic antidepressants, barbiturates, cannabinoids, amphetamines, opiates, and propoxyphene. The technique does not detect drugs at parts per billion (ppb) levels.
Specific drug screening by Enzyme-Linked Immunosorbent Assay (ELISA)	ELISA uses antibodies as a more effective technique compared to EMIT (which it is supplanting). ELISA can detect drugs at ppb levels.
Drug screening by Thin-Layer Chromatography (TLC)	TLC is used a general drug screen in lieu of EMIT and ELISA, which use panels for specific drugs.
General drug screening, identification, and quantification by High-Performance Liquid Chromatography (HPLC)	HPLC is used most commonly in place of TLC, given its greater sophistication and use of computerized compound matching. HPLC can be used as a general screening technique, but also has been configured for specific drug or substance analyses. HPLC is preferred for drugs that decompose in GC/MS injection procedures.
Specific drug identification and quantification by Gas Chromatography linked to Mass Spectrometry (GC/MS)	GC/MS is now the preferred overall technique for analysis of specific drugs. Gas chromatography provides the separation of compounds in wet sample fluids while the mass spectrometry provides the identification and quantification of each analyte using computerized matching compound libraries. Specific separation techniques must be applied to identify specific drugs.

Advances are being made each year in progressively more capable and sensitive analytical equipment and procedures that can be applied to the analysis of substances in autopsy tissues and fluid samples. It is important to apply the most up-to-date techniques when conducting specific analyses for alcohol and drug levels. A number of references are provided at the back of the protocol on the subject of alcohol and drug testing.

III.5 Fire Toxicology

A complete understanding of the cause of a firefighter's death must include some consideration of emergency scene-specific toxicological agents that may have been involved and how they may have interacted with the deceased's biological processes and systems to cause death.

■ For instance, did the inhalation of carbon monoxide result in cardiac ischemia and subsequent cardiac arrest?

■ Did a toxin enter the body through some route other than the respiratory system, such as through dermal exposure, injection, or ingestion?

■ Did protective clothing or SCBA fail to protect the user, or was the user's air supply depleted or otherwise compromised?

These conditions are often accompanied by other injuries which may or may not themselves have caused death, such as crushing forces (trauma) or prolonged exposure to high radiant heat levels (burns).

Firefighters respond to a variety of incidents, each presenting its own unique hazards. Traditionally, most firefighting activity has centered around structural fires. The combustion of wood releases several combustion products into the atmosphere, principally carbon monoxide and other simple hydrocarbons. Structural fires have changed over the past several years because building materials have changed. Roofing, insulation, carpets, paints, and other construction materials all contribute to an ever-growing diversity of chemical products found at fires. The increased use of plastics and other synthetic materials release different kinds of combustion products, many of them highly toxic or carcinogenic. Some examples of fire combustion products:

- carbon monoxide and carbon dioxide;
- inorganic gases (hydrogen sulfide, hydrogen cyanide, nitrogen oxides);
- acid gases (hydrochloric acid, sulfuric acid, nitric acid);
- organic acids (formic acid, acetic acid);
- aldehydes (acrolein, formaldehyde);
- chlorinated compounds (carbon tetrachloride and vinyl chloride);
- hydrocarbons (benzene);
- polynuclear aromatic compounds (PNA); and
- metals (cadmium, chromium).

In addition, chemicals at the site of a fire further contribute to hazardous contaminants in fire smoke. A classic example are PCBs, found in electrical transformers and other equipment, which, when burned, may form dioxin, an acutely deadly substance. Even the normal household will contain cleaning supplies, pesticides, pool chlorine, and other substances that contribute to release of toxic substances at fires. Table 3 lists some common fire smoke contaminants, the sources of these substances, and toxic effects from repeated or high concentration exposure to these chemicals. Table 4 shows chemicals identified in an analysis of fire smoke for several different fires.

Most protective clothing and equipment used by firefighters permits the ready penetration and permeation of toxic chemicals through protective fabrics and components. Since most firefighter protective clothing uses porous fabrics, the chemical vapors or liquids simply penetrate or pass through the pores of the material. Molecules of chemicals can also permeate into the fibers or coatings of clothing materials and can remain in the material for long periods of time, depending on the types of exposure chemical(s) and care given to the clothing. Chemicals that get into the clothing from either means often directly contact the firefighter's skin.

Different areas of the firefighter protective ensemble are likely to demonstrate varying propensities for the absorption or adsorption of chemicals. Any porous fabric material found in the clothing or other items may be contaminated, such as:

- turnout clothing outer shells, moisture barriers, thermal liners, collars, and wristlets;
- station/work uniforms;
- glove shells and liners;
- protective hoods;

■ boot linings;

■ helmet straps; and

■ SCBA straps.

Coated materials such as moisture liners, reflective trim, boot outer materials, a respirator masks are more likely to be affected by permeation. The same is true for hard plastics or resins such as those used in the helmet, SCBA components, and certain turnout clothing hardware.

In addition to liquid or vapor chemical contaminants, a tremendous amount of ash, soot, and other solid matter is released during fires and firefighting activities. This solid matter provides the visible portion of smoke and is the primary cause of residue left on structures and clothing following fires. Soot and ash represent incomplete products of combustion; that is, unburned fuel or agglomerated solids which fail to burn completely during the fire. During combustion, synthetic materials create an increase in the amount of particulate matter, hence the "black" smoke from burning plastics. Since soot particles are very porous, they tend to adsorb other hazardous chemicals. Ash, resins, and other particles from fire smoke can become entrapped within the fibers of clothing or adhere to skin. Accumulation of soot on protective clothing becomes visible as soiled or "dirty" areas. In some cases, these "soils" are made of melted resins or plastics which, in the heat of the fire, become liquid and spread even further throughout the protective clothing. In other cases, many of the particles are too small to see (less than 10 microns) and can penetrate easily into the inner layers of clothing, such as liner and barrier materials, contacting the firefighter's skin.

Table 3. Examples of Fireground Contaminants

Contaminant	Sources	Toxicology
Polychlorinated Biphenyl (PCBs)	Power transformers/capacitors Televisions Air conditioners Carbonless copy paper Hydraulic systems Elevators	• PCBs can produce dioxins that are toxic by inhalation and ingestion. • PCBs also absorb through the skin. • PCBs cause cancer of the liver and pancreas.
Asbestos	Roofing and shingles Acoustic ceiling tiles Sprayed ceilings Old pipe insulation Old octopus-type furnaces Pre-1975 drywall	• Principal hazard is inhalation of fibers (<5 microns length) causes cancer. • Asbestos fibers can be aerosolized from clothing and inspired or and ingested.
Creosote	Power poles Railroad ties Treated wood or buildings Lumber yards Piers and docks	• Creosote is toxic through inhalation and skin absorption. • Causes cancer of skin, prostate, and testicles.
Plastic Decomposition Products • Polycarbonates • Polystyrene • Polyurethane • (PVC)	Electrical insulation Plumbing Furniture Construction materials Insulation and packaging Tools/Toys Automobiles	• Variety of decomposition products including acrylonitrile, hydrogen cyanide, nitrogen oxides, hydrogen chloride, benzene. • Various routes of toxicity through skin absorption, inhalation or ingestion.

Table 4. Specific Chemical Contaminants Identified in Various Fires[15]

Compound	1(K)	1(K)	2(K)	3(O)	4(O)	5(K)	6(K)	6(O)
Furan	X			X				
C_4H_8 isomers	X		X					
Benzene	X	X	X	X	X	X	X	X
Dimethylfuran	X		X					
Methyl methacrylane	X						X	
Toluene	X	X	X				X	
Furfural	X		X					
Xylene	X		X			X		
Styrene	X		X				X	
Pinenes	X		X				X	
Limonene	X						X	
Indane	X		X			X	X	
Methylcyclopentane	X					X		
2,4-Dimethyl-1-pentene						X		
Ethyl benzene						X	X	
C_3-Alkyl benzene						X		
C_4-Alkyl benzene						X	X	
n-Butane							X	
Freon 11							X	
t-Butyl anisole						X	X	
Methyl naphthalene						X	X	

K-knockdown; O-overhaul

Firefighters may be exposed to other particulate hazards. Chemical dusts, lead particles, and asbestos also may be encountered at fires and other responses. For example, though asbestos is principally an inhalation hazard, it can cling to protective clothing and be released when the responder is not wearing his or her SCBA. Similarly, lead and other toxic dusts can fill clothing pores and contaminate the firefighter's skin after the incident.

Firefighters also are subject to exposure to blood or other body fluids containing pathogens, particularly the Human Immunodeficiency Virus (HIV) or Acquired Immunodeficiency Syndrome (AIDS) virus, and Hepatitis B and C viruses. These viruses are extremely small in size and are transmitted by blood or other biological fluids. The risk is high since emergency patient care is a major function of many responses. The extrication of victims from automobile accidents and rescue of injured persons from fires and other incidents all involve the potential for this exposure. Even minute droplets of blood are capable of carrying thousands of virus that potentially can cause infection through mucous membrane contact or nonintact skin. Firefighters also face serious health threats from exposure to existing and nontraditional airborne pathogens that can be encountered in providing medical care or general interface with the public, including tuberculosis, sudden acquired respiratory syndrome (SARS), and more recently avian flu. Though these exposures may not be fatal, they can contribute to firefighter fatalities.

---

[15] Noonan, Gary P., Judith A. Stobbe, Paul Keane, Richard M. Ronk, Scott A. Hendricks, Laurence D. Reed, and Robert L. McCarthy. *Firesmoke: A Field Evaluation of Self-Contained Breathing Apparatus.* NIOSH and U. S. Fire Administration, 1989.

An emerging concern for firefighters and other first responders is the potential lethality from exposure to chemical, biological, radiological, nuclear, and explosive (CBRNE) hazards. These hazards may take the form of chemical warfare agents, toxic industrial chemicals, biological agents that are both liquid and airborne, ionizing radiation, nuclear material, and explosives. Terrorism has become a real threat to firefighters and is likely to cause multiple casualties, including firefighters. Under the circumstances of a terrorism event, special provisions will be needed for the handling of first responder and civilian deaths.

III.6 Burns

Firefighters encounter flames, high heat, physical obstacles, and a number of other hazards in carrying out their response duties. Each hazard serves as an individual stressor on the firefighter that given its relative intensity, length of exposure, and the degree to which protection is provided, creates specific risks of injury, disease, or death.

Particularly relevant to this report is structural firefighting, which the NFPA defines as "the activities of rescue, fire suppression, and property conservation in buildings, enclosed structures, vehicles, marine vessels, or like properties that are involved in a fire or emergency situation." Structural firefighting is likely to expose firefighters to a range of thermal conditions when responding to a fire. While several researchers have attempted to classify these conditions, one system is shown in Figure 1, where the fireground is characterized in terms of level of thermal radiation (expressed in cal/cm²s) and the air temperature (expressed in degrees Celsius and degrees Fahrenheit).[16,17] Three possible structural firefighting situations are illustrated in this figure and are described below:

- ■ The **Routine** region describes conditions where one or two objects, such as a bed or waste basket, are burning in a room. The thermal radiation and the air temperatures are virtually the same as those encountered on a hot summer day. As shown in Figure 1, **Routine** conditions are accompanied by a thermal radiation range of 0.025 to 0.05 cal/cm2s and by air temperatures ranging from 68 to 140 °F (20 to 60 °C). Protective clothing for firefighters typically provides protection under these conditions, but excessive exposure times may create a burn injury situation.

- ■ The **Ordinary** region describes temperatures encountered in fighting a more serious fire or being next to a "flashover" room. **Ordinary** conditions are defined by a thermal range of 0.05 to 0.6 cal/cm2s, representing an air temperature range of 140 to 571 °F (300 °C). Under these conditions, protective clothing may allow sufficient time to extinguish the fire or to fight the fire until the nominal air supply is exhausted (usually less than 30 minutes).

- ■ The **Emergency** region describes conditions in a severe and unusual exposure, such as those caused inside a "flashover" room or the firefighter being next to a flame front. In **Emergency** conditions, the thermal load exceeds 0.3 cal/cm2s and temperatures exceed 571 °F. In such conditions, the function of firefighters' clothing and equipment is simply to provide protection during the short time needed for an escape without serious injury.

---

[16] Abbott, N. J. and S. Schulman. "Protection from Fire: Nonflammable Fabrics and Coatings." *Journal of Coated Fabrics*, Vol. 6, July 1976, pp. 48-64.

[17] Utech, H.P. "High Temperatures vs. Fire Equipment." *International Fire Chief*, Vol. 39, 1973, pp. 26-27.

Figure 6. Classification of Fireground Conditions

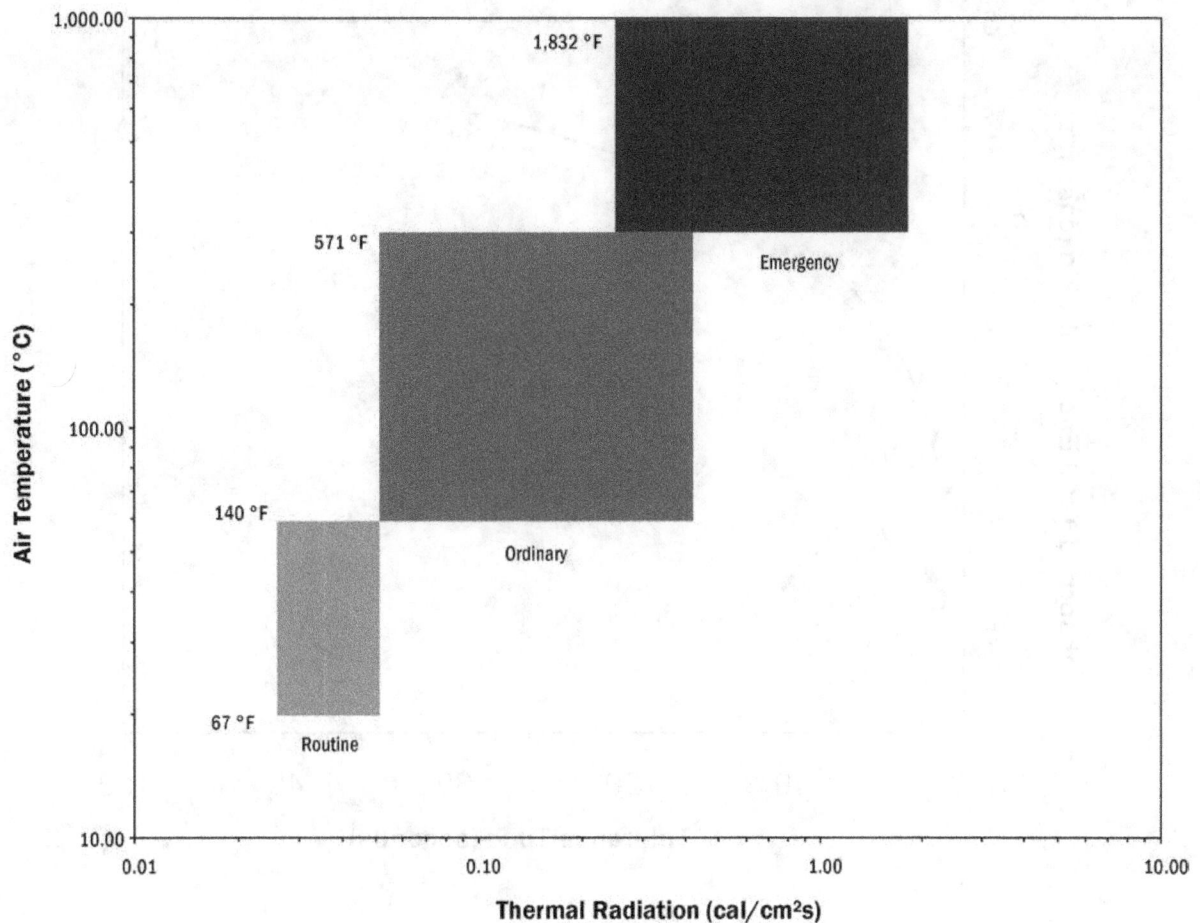

Burns occur as the consequence of heat transfer to the skin at a rate greater than the body's ability to dissipate that heat. The occurrence of a burn injury can be predicted by plotting the heat energy versus time, as shown in Figure 7. This relationship shows that burns occur very quickly for high levels of heat energy exposure, while exposures to lower heat energies require a greater amount of time to cause a burn injury. Thus, burn injury is a function of both the intensity of heat exposure and the length of the exposure.

Increasing amounts of stored heat energy in the skin cause progressively greater damage. The defined degrees of burn injury distinguish the levels (depth) of skin damage and how permanent the damage becomes. For example, third-degree burns, also referred to as full-thickness burns, involve damage to the entire skin thickness and are considered irreversible (complete healing is not possible). It is also important to point out that burn injuries may occur under portions of protective clothing and equipment that show no damage.

Figure 7. Relationship of Heat Energy and Time to Burn Injury

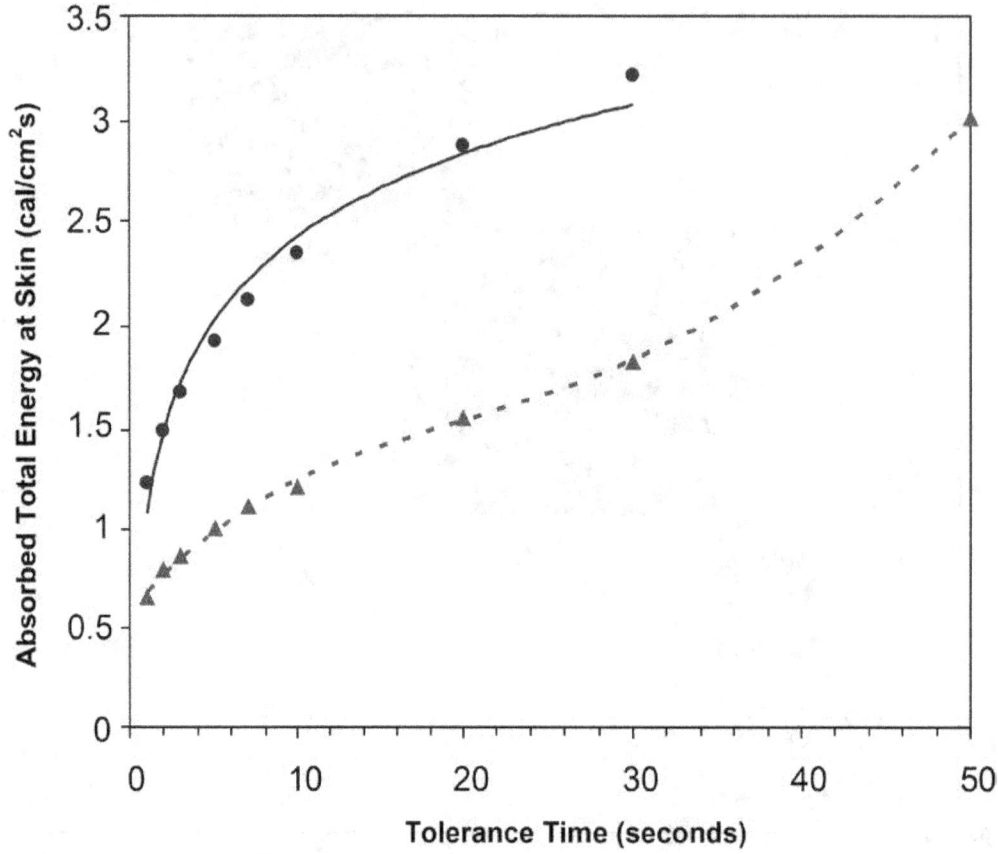

III.7 Personal Protective Equipment

Detailed knowledge of the manner of death requires, among other things, an evaluation of the performance of the firefighter's personal protective equipment (PPE), which includes protective clothing and breathing apparatus. There is voluminous anecdotal evidence that failure to use proper protective equipment has been responsible for many firefighter injuries, illnesses, and deaths.

Typically, firefighter protective ensembles consist of several elements of clothing and equipment that are worn together to provide protection against fireground hazards:

■ SCBA;

■ protective coat and pants;

■ protective helmet;

■ protective hood;

■ protective gloves;

■ protective footwear; and

■ PASS (may be integrated with SCBA)

An example protective ensemble for structural firefighting is shown in Figure 8. Additional PPE information is provided in Appendix C. Other types of specialized ensembles are worn by firefighters for different applications. These include specialized ensembles for emergency medical operations, hazardous materials incidents, and technical rescue events.

Figure 8. Typical Structural Firefighting Ensemble (Courtesy of Morning Pride Manufacturing)

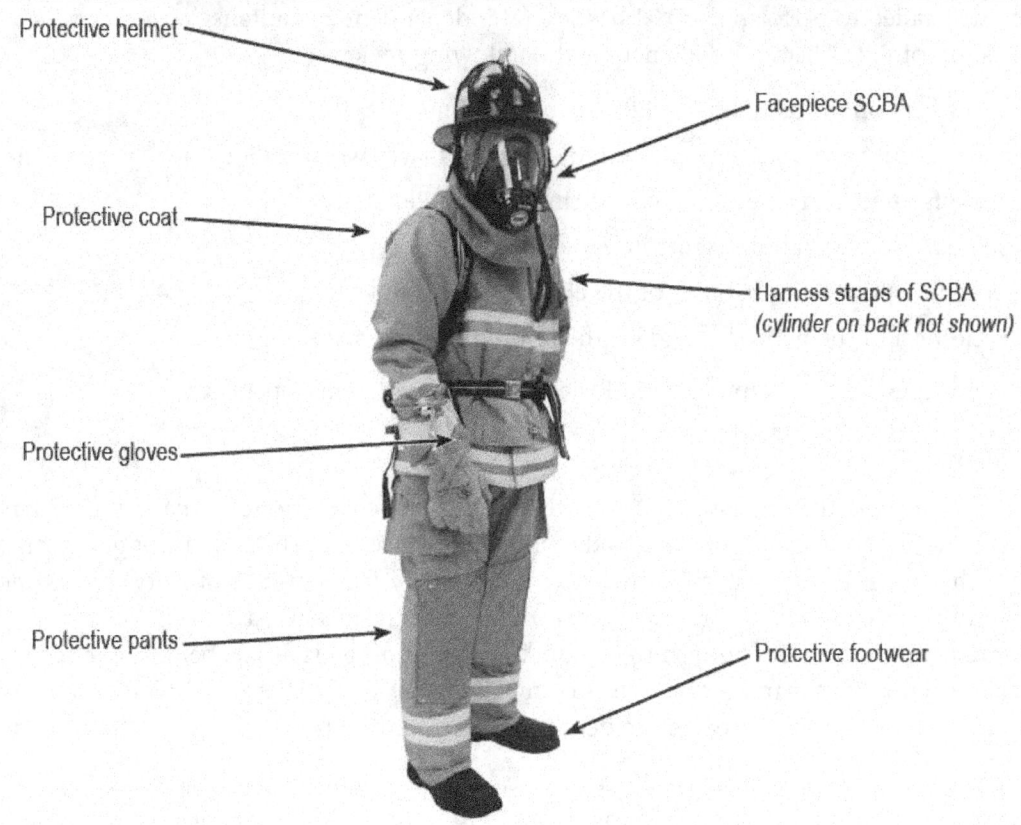

Protective helmet

Facepiece SCBA

Protective coat

Harness straps of SCBA
(cylinder on back not shown)

Protective gloves

Protective pants

Protective footwear

In order to provide intended protection, the ensemble elements must be chosen to work together without interference and must be properly sized and worn consistent with manufacturer instructions. Even when these instructions are followed, it still is possible to overwhelm the protective qualities of the firefighter PPE when fireground conditions exceed the designed capabilities of the clothing and equipment. In addition, the burden of insulative clothing under hot working conditions also creates stressors on the firefighter's body that can contribute to illness and death.

The use of SCBA has reduced significantly the number of firefighter injuries and deaths that are attributable to smoke inhalation. While thermal and respiratory injuries remain a concern in cases of firefighter autopsies, the widespread use of SCBA has introduced new considerations into the evaluation of these injuries. For example, knowing that a firefighter's death was the result of inhalation of combustion products, when the firefighter was using an SCBA, would indicate the need to fully evaluate the performance of the SCBA. This investigation can be conducted by the NIOSH, which is responsible for the certification of this equipment. Reviews are particularly important to ascertain the correct functioning of the equipment and the amount of service air remaining in the cylinder. Obtaining this information may not always be possible when the SCBA is destroyed in the fire. Regardless, it is important to check the service life of the SCBA as part of any investigation, particularly where there is a question of asphyxiation. The actual service life provided by an SCBA is always significantly less than the rated service life, since firefighters can exhaust the air supply through more rapid breathing due to physical activity and stress. Some instances also may be able to compromise the positive pressure of the facepiece, permitting the infiltration of outside contaminants.

Experts may need to be consulted to determine how a firefighter's protective clothing and equipment performed or failed to perform. NIOSH has several independent consultants who are available to assist in the evaluation of PPE. Checks of PPE include the following reviews:

■ the identification on each item of PPE that was worn;

■ the identification of each personal item of clothing that is worn underneath the protective clothing;

■ the manufacturer and date of manufacturing of each PPE item;

■ the standard to which the specific item of PPE complies;

■ any specific options or attributes of the clothing item; and

■ the condition of the item as found on the injured or deceased firefighter.

Specific guidelines for the examination of PPE items are provided in Appendix C.

III.8 Non-Line-of-Duty Deaths

Because of their repetitive exposure to toxic environments and carcinogens, many firefighters are concerned that they are at a higher risk to die prematurely, particularly as their longevity on the job increases. The causes of firefighter deaths that occur off-duty (or non-line-of-duty) sometimes can be attributed to one exposure or to a series of exposures to toxins. There have been some major, well-documented exposures of firefighters to certain known carcinogens. It has been suggested, for instance, that fires in occupancies manufacturing or storing chemicals in Elizabeth, New Jersey, and Fort Lauderdale, Florida, are responsible for increased incidence of cancer among the firefighters who fought these blazes.

In one of those examples, as many as 29 cases of cancer, including 19 cancer deaths, have occurred among the approximately 100 firefighters who fought a fire in 1968 at the Everglades Fertilizer Plant in Fort Lauderdale, Florida. All but one of these cases was diagnosed after the firefighter had retired or resigned from the fire department. This case has prompted the NIOSH to initiate an epidemiological study of firefighters involved in the Everglades fire.

More recently, additional medical and industry reviews have found the incidence of firefighter cancers to be elevated as compared to other occupations.

■ A study of Seattle firefighters shows excess mortality from diseases of a priori concern, such as lung cancer, nonmalignant respiratory disease, and cardiovascular disease.[18]

■ An analysis of firefighter deaths in the Boston area indicated a 3-to-1 increase in firefighter cancers when compared to the general population.[19]

---

[18] Heyer, Nicholas, Noel S. Weiss, Paul Demers, and Linda Rosentock. "Cohort Mortality Study of Seattle Fire Fighters: 1945-1983." *American Journal of Industrial Medicine*, Vol. 17, 1990, pp. 493-504.

[19] Musk, A. William, John M. Peters, and David H. Wegman. "Lung Function in Fire Fighters, I: A Three Year Follow-Up of Active Subjects." *American Journal of Public Health*, Vol. 67(7), 1977, pp. 86-89.

■ In Los Angeles, cancer mortality among surviving firefighters is elevated for all lung and circulatory system cancers.[20]

■ A study of firefighter mortality as compared to police officers in three northwestern U.S. cities showed excesses of brain tumors, and lymphatic and hematopoietic cancers.[21]

■ A detailed mortality study in Toronto showed firefighters to have statistically significant excesses of brain, central nervous system, and other cancers.[22]

A more recent study has documented the risk of increased rates of multiple myeloma, non-Hodgkin lymphoma, and testicular cancer for firefighters based on a review of 32 different studies related to firefighter cancer risks.[23]

It can be very difficult to directly attribute a non-line-of-duty death to a line-of-duty exposure, especially if the exposure occurred years before the death. Comprehensive autopsies of firefighters whose death may have been caused by a line-of-duty exposure could help establish a better understanding of the relationship between exposures and premature deaths; however, this will require much better data be obtained and maintained than is currently the norm.

Many fire departments have mandated physical requirements and medical examinations for firefighters. Regular medical exams and physical testing can track a firefighter's physical and medical status from hire to retirement, and can serve as a baseline against which to compare, especially after an incident or series of incidents where a firefighter may be concerned that an exposure has jeopardized his or her health.

Records of exposures to particular toxins should be kept by the fire department along with the medical records. Such documentation would be valuable in determining whether an exposure led to medical problems, or whether a non-line-of-duty death is related to firefighting or other emergency or occupational activities.

The firefighter autopsy protocol is primarily intended to be applicable to line-of-duty deaths; however, it would also be appropriate for non-line-of-duty deaths where an occupational factor is suspected to be involved in the cause of death. For most firefighter deaths that are not duty-related or that involve former firefighters, existing clinical autopsy procedures consistent with the individual's medical history should be appropriate. The USFA Firefighter Autopsy Protocol has been designed to uncover pertinent forensic information consistent with the distinct occupational aspects of firefighting.

III.9 Firefighter Health

Several studies have looked at the frequency of premature death rates among active and retired firefighters. Rubin has described the relationships between the hazards of fire suppression and the risk of premature

---

[20] Lewis, S.S., H.R. Bierman, and M.R. Faith. "Cancer Mortality Among Los Angeles City Fire Fighters." Published Report Submitted to the Los Angeles Fire Department, Dec., 1982.

[21] Demers, Paul A., Nicholas J. Heyer, and Linda Rosenstock. "Mortality Among Firefighters from Three Northwestern United States Cities." *British Journal of Medicine*, 1992, 49: 664-670.

[22] L'Abbe, Kristan A. and George A. Tomlinson. "Fire Fighters in Metropolitan Toronto: Summary of the Mortality Study." *Industrial Standards Panel*, Toronto, Ontario, Canada, 1994.

[23] LeMasters, Grace K. et al. "Cancer Risk Among Firefighters: A Review and Meta-Analysis of 32 Studies." *Journal of Environmental Medicine*, Vol. 48, No. 11, November 2006, pp. 1189-1202.

death from heart disease or cancer as "Firefighter's Disease."[24] He notes that relatively little research has been conducted on firefighter mortality and morbidity or the medical treatment of firefighters.

Rubin proposes that a concern for firefighter health should begin with prevention. He suggests that diet, lack of exercise, and lifestyle may be as responsible for premature firefighter deaths as any job-related exposure. The relationships of lifestyle, exercise, and diet with firefighter mortality appear to be more than just conjecture. Epidemiological studies have demonstrated that firefighters are much less likely than the general population to die from natural causes at a given age, early in their careers, because they must be healthier than the average person to pass the rigorous health and fitness standards in order to be hired or approved for volunteer duty. The death rate for firefighters catches up with the rest of the population by their retirement age, which suggests that the so-called "healthy worker effect" diminishes with time, especially if the individuals do not take care of themselves. This takes into account the factor that firefighters tend to retire at a younger age than the general population.

The USFA[25] describes a number of resources that address the implementation of firefighter health programs. These include the IAFF/International Association of Fire Chiefs (IAFC) "Fire Service Joint Labor Management Wellness-Fitness Initiative;" the USFA/National Volunteer Fire Council (NVFC) "Health and Wellness Guide for the Volunteer Fire Service"; the NVFC "Heart-Healthy Firefighter Program"; the IAFF/IAFC "Candidate Physical Ability Test" for recruits; and the 16 Life Safety Initiatives from the National Fallen Firefighter Foundatin's (NFFF's) "Everyone Goes Home" program.

---

[24] Rubin, D.L. "Firefighters' Disease.," *Firehouse Magazine*, Jan. 1992, pp. 40-43. U.S. Bureau of the Census. 1991. *Statistical Abstract of the United States*, 111th ed. Washington, D.C.: U.S. Government Printing Office.

[25] Fire in the United States: 1992-2001, Chapter 5, p. 153.

IV. PUBLIC SAFETY OFFICER BENEFITS (PSOB) PROGRAM

IV.1 Summary of the PSOB Law[26]

The Public Safety Officers Benefits Act, (42 USC 3796, Public Law 94-430) became law on September 29, 1976. The legislation provided for a $50,000 death benefit for firefighters (paid and volunteer) and law enforcement officers who died in the line of duty (emergency or non-emergency) from a traumatic injury. On December 15, 2003, the Act was amended (Hometown Heroes Act) to cover deaths from heart attack and stroke occurring in the line of duty. The Act does not cover deaths resulting from occupational illness or pulmonary disease unless a traumatic injury is a substantial factor to the death. On August 10, 2006, new regulations for the PSOB were issued that incorporated all prior amendments to the original regulations and adds the regulations for the Hometown Heroes Act.

On November 11, 1988, the benefit was increased from $50,000 to $100,000 and made retroactive to June 1, 1988. The dependency test for parent(s) was eliminated. Additionally, it provided that, on October 1, 1988, and every year thereafter, the benefit would be increased to reflect any increase in the consumer price index. On October 26, 2001, as part of the Patriot Act of 2001, the benefit was increased to $250,000 and made retroactive to January 1, 2001.

The Act did exclude Federal firefighters; however on October 12, 1984, the Act was amended to correct this exclusion. Likewise, on October 15, 1986; public-sector EMS personnel also were amended into the coverage of the Act. On June 25, 2002, the Act was amended by the enactment of the Mychal Judge Police and Fire Chaplains Safety Officers Benefit Act, which now allows coverage of fire chaplains under the Act and authorizes all beneficiaries of fallen firefighters, not just parents, spouses; or children to receive the Federal compensation. The legislation, named after the FDNY Chaplain Father Judge, was proposed after it was discovered that 10 public safety officers who gave their lives on Sept. 11 would not be eligible for death benefits because they did not have any surviving immediate family.

On August 10, 2006, new regulations for administration of all PSOB benefits were issued that incorporated all prior amendments to the original regulations and added the provisions of the Hometown Heroes Act (see below). These new regulations address the PSOB Act and regulations in five parts:

1. The first part of this document describes the structure and background of the PSOB Program and aspects of the history of its administration.

---

[26] From Summary of the Federal (U.S.) Benefits for Public Safety Officers PSOB and PSOEA Programs, IAFF Division of Occupational Health Safety and Medicine, www.iaff.org

2. The second part covers the recent changes to the PSOB Act contained in Public Law 109–162, which provided a number of clarifying and conforming changes to the PSOB Act. New definitions included the term "member of a rescue squad or ambulance crew," which is now defined as "an officially recognized or designated public employee member of a rescue squad or ambulance crew." It also amended the PSOB Act to ensure that the pre-existing statutory limitation on payments to noncivilians referred to the individual who was injured or killed, and not to any potential beneficiaries. Finally, this legislation amended certain provisions of the PSOB Act regarding designation of beneficiaries when the officer dies without a spouse or eligible children and removed the need for a 1-year waiting period to ensure payment to the beneficiary of the officer's "most recently executed life insurance policy."

3. The third part addresses the comments received by the Bureau of Justice Assistance (BJA) that relate to the proposed provisions implementing the Hometown Heroes Act, and explains the changes being made in the final rule.

4. The fourth part is a specific discussion of the terms "line of duty" and "authorized commuting."

5. The last part addresses the remainder of the comments in a section-by-section analysis, indicating where changes to provisions were made, or where the BJA determined no changes were necessary.

IV.2 Summary of the Heart/Stroke Amendment[27]

The Hometown Heroes Survivors Benefits Act of 2003 (HHA) amends the PSOB Act and was signed into Law on December 15, 2003. If a public safety officer dies as a direct and proximate result of a heart attack or stroke, that officer shall be presumed to have died as the direct and proximate result of a personal injury sustained in the line of duty unless such presumption is not overcome by competent medical evidence to the contrary.

The law requires that the officer, while on duty engaged in a situation, and such engagement involved nonroutine stressful or strenuous physical law enforcement, fire suppression, rescue, hazardous material response, emergency medical services (EMS), prison security, disaster relief, or other emergency response activity; or participated in a training exercise, and such participation involved nonroutine stressful or strenuous physical activity. Any claim for nonroutine stressful or strenuous physical activities will be excluded if such actions are of a clerical, administrative, or nonmanual nature.

Further, the law requires that the officer must have died as a result of a heart attack or stroke suffered

■ while engaging or participating in such activity as described above;

■ while still on that duty after so engaging or participating in such an activity; or

■ not later than 24 hours after so engaging or participating in such an activity.

The HHA provision only covers deaths occurring on or after December 15, 2003. The HHA is not retroactive, and therefore it does not apply to deaths that occurred before the aforementioned date.

---

[27] From Summary of the Federal (U.S.) Benefits for Public Safety Officers PSOB and PSOEA Programs, IAFF Division of Occupational Health Safety and Medicine, *www.iaff.org*

IV.3 Useful Web sites

The following Web sites provide useful information either directly or indirectly related to the Public Service Officer Benefits Program:

Department of Justice Public Service Officer Benefits	*https://www.psob.gov/*
National Fallen Firefighters Foundation	*http://www.firehero.org/*
U.S. Fire Administration (firefighter fatality page)	*http://www.usfa.dhs.gov/fireservice/fatalities/index.shtm*
National Institute for Occupational Safety and Health (Fire Fighter Fatality Investigation and Prevention Program)	*http://www.cdc.gov/niosh/fire/*
International Association of Fire Fighters (line-of-duty deaths)	*http://www.iaff.org/HS/LODD/index.html*
International Association of Fire Chiefs (Near Miss reporting system)	*http://www.iafc.org/displaycommon.cfm?an=1&subarticlenbr=328#nearmiss*
National Volunteer Fire Council (in the line of duty)	*http://nvfc.org/index.php?id=657*

Selected Bibliography

Determination of Alcohol Levels During Autopsy (*abstract provided where available*)

Barillo DJ, Rush BF Jr, Goode R, Lin RL, Freda A, Anderson EJ Jr
Am Surg.1986 Dec;52(12):641-5.
Is ethanol the unknown toxin in smoke inhalation injury?

Of the 12,000 fire-related deaths occurring annually in the United States, it is estimated that 60 to 80 per cent are due to smoke inhalation. Plastic and synthetic materials which have been introduced in home construction and furnishings produce a more toxic smoke when burned. Efforts to identify a "supertoxin" in this smoke have been unsuccessful to date. An alternative approach is to examine why victims are unable to escape, and become exposed to smoke for lethal periods of time. The authors examined the circumstances of death in 39 fire victims (27 adults, 12 children) over a 25-month period. Detailed examination of the fire scene, autopsy studies, and toxicologic analysis were carried out. Position of the victim, and escape efforts were noted. Carbon monoxide was elevated in all victims, with "lethal" levels (= greater than 50%) in 21/39 victims. Cyanide was detected in 24/29 victims, but none had lethal (3 mg/L) levels present. Ethanol was detected in 21/26 adults (80%) and 0/12 children (0%). 18/26 adult victims had ethanol levels above the statutory level of intoxication (10 mg%). Victims found in bed (no escape attempt) had a mean blood ethanol level of 268 mg%, compared with a mean level of 88 mg% in those victims found near an exit (P = .006). Ethanol intoxication significantly impairs the ability to escape from fire and smoke and is a contributory factor in smoke-related mortality.

Bonnichsen R, Moller M, Maehly AC.
Zacchia.1970 Apr-Jun;6(2):219-25.
How reliable are post-mortem alcohol determinations?

Brown GA, Neylan D, Reynolds WJ, Smalldon KW.
Anal Chim Acta.1973 Sep;66(2):271-83.
The stability of ethanol in stored blood. I. Important variables and interpretation of results.

Buchsbaum RM, Adelson L, Sunshine I.
Cuyahoga County Coroner's Office, Cleveland, OH 44106.
Forensic Sci Int.1989 Jun;41(3):237-43.
A comparison of post-mortem ethanol levels obtained from blood and subdural specimens.

Post-mortem subdural ethanol levels have been proposed as a useful test in certain forensic cases involving head trauma, particularly when the time interval from injury to death may have caused a lowering of the blood ethanol concentration to insignificant or undetectable levels. This study of 75 autopsied persons from whom both blood and subdural ethanol levels were obtained, shows the usefulness of the subdural ethanol level, especially where there is a prolonged or unknown post-traumatic time interval. Use of such a test is recommended in these situations.

Budd RD.
J Chromatogr. 1982 Dec 3;252:315-8.
Ethanol levels in postmortem body fluids.

Chao TC, Lo DS.
Institute of Science and Forensic Medicine, Singapore.
Am J Forensic Med Pathol. 1993 Dec;14(4):303-8.
Relationship between postmortem blood and vitreous humor ethanol levels.

The relationship between the blood to vitreous humor ethanol ratios (B/V) and the corresponding urine to blood ethanol ratios (U/B) of 200 postmortem cases were found to be bimodal in nature. Using the U/B ratio of 1.20 as a demarcation below which early absorption prevails, the results in the early absorption phase gave an average B/V ratio of 1.29, a range from 0.71 to 3.71, and a relatively large standard deviation of 0.57, whereas the results in the other phases (late absorption and elimination) gave an average B/V ratio of 0.89, a spread from 0.32 to 1.28, and a standard deviation of 0.19. It would appear that the blood ethanol levels can be estimated using B = 1.29 V for early absorption phase cases and B = 0.89 V for cases in subsequent phases. The former relationship would underestimate the blood ethanol levels in cases of very early absorption phase and the later overestimate the levels of late elimination cases. The ethanol distribution results in cases of fatal road traffic accidents and suicides by falling, in which 69% of the deceased sustained some form of head injury, were found to be similar to those of other postmortem cases. The observations reflect that vitreous humor, being reasonably protected, is likely to survive certain traumatic deaths and be available for postmortem ethanol investigation. The U/B ethanol ratios recorded in this work had an average of 1.29, a range from 0.19 to 5.19, and a standard deviation of 0.48.

Coe JI, Sherman RE.
J Forensic Sci. 1970 Apr;15(2):185-90.
Comparative study of postmortem vitreous humor and blood alcohol.

de Lima IV, Midio AF.
University of Sao Paulo, Medicolegal Institute, College of Pharmaceutical Sciences, Brazil.
Forensic Sci Int. 1999 Dec 20;106(3):157-62.
Origin of blood ethanol in decomposed bodies.

Problems related to blood contamination by other postmortem fluids in decomposed bodies (DB) make the interpretation of medicolegal blood alcohol levels (B EtOH) a very difficult task. So the aim of this paper is to show the utilization of vitreous humor (VH) as the biological fluid for an unequivocal determination of ethanol origin in DB for forensic purposes. Alcohol was determined in VH, blood (chest fluid-CF) and urine (Ur) collected from 27 DB in different states of putrefaction. A simple head-space gas-chromatographic method was used. In fifteen cases alcohol was found to be of endogenous production due to its absence in VH. In the twelve remainders, alcohol was detected in

VH and CF in an atypical distribution. Examining the reliable scene and historical information together with the analytical data, ethanol origin in these cases was classified: endogenous production (3 cases), ingested (2 cases), both (2 cases), contaminated plus endogenous production (3 cases) and unable to determine (2 cases). According to the results obtained it was possible to conclude that alcohol analysis in VH is fundamental for determining the origin of ethanol detected in CF of DB.

Hardin GG.
Forensic Science Laboratory, Minnesota Department of Public Safety Bureau of Criminal Apprehension, St. Paul 55104, USA.
J Forensic Sci.2002 Mar;47(2):402-3.
Comment in:
   J Forensic Sci. 2002 Nov;47(6):1405; author reply 1405.
Postmortem blood and vitreous humor ethanol concentrations in a victim of a fatal motor vehicle crash.

A 20-year-old male was found on the passenger side of a small car after a collision with a semi-trailer truck. Postmortem blood, collected from the chest cavity, and vitreous humor samples were collected following harvesting of the heart and bones. Gas chromatographic analysis revealed a blood ethanol concentration of 0.32 g/dL and a vitreous humor ethanol concentration of 0.09 g/dL. The stomach was intact and full of fluid and food, but its contents were not collected. Possible explanations for the large difference between the two results include diffusion of ethanol from the stomach into the chest cavity, contamination of the blood sample prior to collection, and ingestion of a large quantity of ethanol shortly before death. This case demonstrates the importance of proper quality assurance procedures in collecting postmortem specimens and of collecting a vitreous humor sample for ethanol analysis in postmortem toxicology cases.

Heise HA.
Rocky Mt Med J.1968 Jun;65(6):39-44.
Alcohol and sudden death--importance of testing several body fluids.

Ito A, Moriya F, Ishizu H.
Department of Legal Medicine, Okayama University Medical School, Japan.
Acta Med Okayama.1998 Feb;52(1):1-8.
Estimating the time between drinking and death from tissue distribution patterns of ethanol.

To establish a method for estimating the time between the last consumption of alcohol and death, we examined the ethanol levels in body fluids and tissues of rats that had been orally administered 1 g/kg ethanol. We observed the following relationships between ethanol levels in the cardiac blood (blood in the heart itself), vitreous humor, and urine: cardiac blood > vitreous humor > urine at 10 min (early absorption stage); vitreous humor > cardiac blood > urine from 20 to 50 min (late absorption stage); vitreous humor > urine > cardiac blood from 60 to 120 min (distribution stage); and urine > vitreous humor > cardiac blood at 180 min (excretion stage). It was also observed that, in cases of death immediately following drinking, ethanol levels in the stomach contents are very high, and the following ratios of ethanol levels were observed: skeletal muscle to cardiac blood--less than 1; liver to cardiac blood--around 1. buccal mucosa to cardiac blood-greater than 1. These ratios at equilibrium after drinking were around 1, lower than 1 and around 1, respectively. We also measured alcohol levels in the cardiac blood, urine, vitreous humor and stomach contents of nine cadavers who had consumed alcohol prior to death. The relationships between the time since last consumption of alcohol and relative ethanol levels in these specimens were in good accordance with the results of the animal experiments.

A W Jones and P Holmgren
J. Clin. Pathol., Sep 2001; 54: 699 - 702.
Uncertainty in estimating blood ethanol concentrations by analysis of vitreous humour

Aims—To determine the concentrations of ethanol in femoral venous blood (FVB) and vitreous humour (VH) obtained during forensic necropsies. The ratios of ethanol concentrations in VH and FVB, the reference interval, and the associated confidence limits were calculated to provide information about the uncertainty in estimating FVB ethanol concentrations indirectly from that measured in VH. *Methods*—Ethanol concentrations were determined in specimens of FVB and VH obtained from 706 forensic necropsies. The specimens were analysed in duplicate by headspace gas chromatography (HS-GC), with a precision (coefficient of variation) of 1.5% at a mean ethanol concentration of 500 mg/litre. The limit of detection of ethanol in body fluids by HS-GC in routine casework was 100 mg/litre. *Results*—In 34 instances, ethanol was present in VH at a mean concentration of 154 mg/litre, whereas the FVB ethanol concentration was reported as negative (< 100 mg/litre). These cases were excluded from the statistical analysis. The concentration of ethanol in FVB was higher than in VH in 93 instances, with a mean difference of 160 mg/litre (range 0 to 900). The mean concentration of ethanol in FVB (n = 672) was 1340 mg/litre (SD, 990) compared with 1580 mg/litre (SD, 1190) in VH. The arithmetic mean VH/FVB ratio of ethanol was 1.19 (SD, 0.285) and the 95% range was 0.63 to 1.75. The mean and SD of the differences (log VH - log FVB) was 0.063 (SD, 0.109), which gives 95% limits of agreement (LOA) from -0.149 to 0.276. Transforming back to the original scale of measurement gives a geometric mean VH/FVB ratio of 1.16 and 95% LOA from 0.71 to 1.89. These parametric estimates are in good agreement, with a median VH/FVB ratio of 1.18 and 2.5th and 97.5th centiles of 0.63 and 1.92. *Conclusions*—The ethanol distribution ratios (VH/FVB) show wide variation and this calls for caution when results of analysing VH at necropsy are used to estimate the concentration in FVB. Dividing the ethanol concentration in VH by 2.0 would provide a very conservative estimate of the ethanol content in FVB, being less than the true value, with a high degree of confidence.

Leahy MS, Farber ER, Meadows TR.
J Forensic Sci. 1968 Oct;13(4):498-502.
Quantitation of ethyl alcohol in the postmortem vitreous humor.

Kaye S.
Am J Clin Pathol. 1980 Nov;74(5):743-6.
The collection and handling of the blood alcohol specimen.
Text Available @ http://rcm-medicine.upr.clu.edu/publications/sidney_kaye/the-collection-and-handling-of-blood.htm

Proper collection, handling, and storage of the blood alcohol specimen are essential in medicolegal cases involving the question of sobriety. A standard operating procedure is necessary to ensure maximum reliability. Comments are offered on the advantages of using blood specimens in preference to urine or tissue specimens. The use of a conversion factor to obtain a calculated "presumed blood level" can be dangerous. Cautions and suggestions are offered regarding how and from where the blood should be obtained from a living person and during an autopsy. There are certain time limitations for storage of these blood-alcohol specimens. Each laboratory must establish its own limits for reliable storage, given the conditions in that laboratory. Unexpected and confusing results can lead to an erroneous interpretation if history, circumstances, type of injury, and survival time are not all carefully considered. Several possibilities for error in judgment are discussed.

Kaye S, Cardona E.
Am J Clin Pathol.1969 Nov;52(5):577-84.
Errors of converting a urine alcohol value into a blood alcohol level.
Text Available @ *http://rcm-medicine.upr.clu.edu/publications/sidney_kaye/Error-of-Converting.htm*

A blood alcohol determination is one of the more frequently requested analyses in a toxicology or forensic chemical laboratory. There are many reliable methods for determining the concentration of alcohol in the blood. It is the purpose of this communication to show that it is, however, not reliable to determine the concentration of alcohol in the urine and report as a blood alcohol level. This is not reliable even using the best of average factors of equivalence. These factors used are an average of many determinations, some of which show very wide ranges from the mean. Random specimens of urine and blood were collected from 148 cases examined for alcohol content. An average urine-blood alcohol ratio of 1.28: 1, with a range of 0.21 to 2.66, was obtained. The blood alcohol level was calculated in each case from the corresponding urine alcohol determination by means of the average ratio obtained from our data. In 32 (21.5%) of the cases, the blood figures calculated from the urine value exceeded the actual level determined in blood. In 51 cases (34.5%) the calculated blood alcohol concentration was below the determined value. In 65 cases (44%) the values corresponded. This procedure was repeated using the conversion factor (1.25: 1) employed in some communities. In this instance, the calculated blood alcohol concentration exceeded the actual value in 39 cases (26.5%). In 49 cases (33%) the calculated value was below the observed level, and in 60 cases (40.5%) the values corresponded. In view of the wide ranges in the individual urine-blood alcohol ratios found in most published reports, we find it hard to understand how so many investigators can conclude that it is satisfactory procedure to calculate the alcoholic content of blood, to the second decimal place, from a selected specimen of urine. Our data clearly confirm what other investigators[2, 7 – 10] have claimed: that the relationship (ratio-range) between the concentrations of alcohol in urine and in blood may vary widely. This renders it unreliable to use an average conversion factor in medicolegal cases.

Kuroda N, Williams K, Pounder DJ.
Department of Legal Medicine, Keio University, Tokyo, Japan.
Am J Forensic Med Pathol.1995 Sep;16(3):219-22.
Estimating blood alcohol from urinary alcohol at autopsy.

Urine alcohol concentration (UAC) and blood alcohol concentration (BAC) measured by gas chromatography were available from 435 medicolegal autopsies. Simple linear regression with BAC as outcome variable and UAC as predictor variable (range, 3-587 mg%) gave the regression equation BAC = -5.6 + 0.811UAC with 95% prediction interval +/- 0.026 square root of [9465804 + (UAC-213.3)2] and 99% prediction interval +/- 0.034 square root of [9465804 + (UAC-213.3)2]. The standard error of the slope was 0.013 and the 95% confidence interval for the slope 0.785-0.837. In practice, a BAC of 80 mg% is predicted with 95% certainty by a UAC of 204 mg% and similarly a BAC of 150% by a UAC of 291 mg%. The prediction interval is too wide to be helpful in the assessment of an individual case fatality. The UAC is useful in corroborating but not in predicting BAC.

Mackey-Bojack S, Kloss J, Apple F.
Hennepin County Medical Center, Clinical Laboratories, Minneapolis, Minnesota 55415, USA.
J Anal Toxicol. 2000 Jan-Feb;24(1):59-65
Cocaine, cocaine metabolite, and ethanol concentrations in postmortem blood and vitreous humor.

The use of postmortem cocaine and metabolite concentrations is a complex subject. This study was undertaken to determine (1) the usefulness of vitreous humor as a specimen, compared with blood, to quantitate cocaine and cocaine metabolites; (2) whether there is a preferential site of disposition for cocaethylene between vitreous humor and blood; and (3) if the presence of cocaethylene influences the concentration of benzoylecgonine in postmortem specimens. Cocaine, benzoylecgonine, and cocaethylene were quantitated in blood and vitreous humor by gas chromatography-mass spectrometry, and ethanol was quantitated by gas chromatography in 62 medical examiner cases. No differences were found between mean concentrations of vitreous cocaine 0.613 mg/L (standard deviation [SD] 0.994 mg/L), cocaethylene 0.027 mg/L (SD 0.59 mg/L), and ethanol 0.092 g/dL (SD 0.13 g/dL) compared to blood cocaine 0.489 mg/L (SD 1.204 mg/L), cocaethylene 0.022 mg/L (SD 0.055 mg/L), and ethanol 0.058 g/dL (SD 0.91 g/dL), respectively. However, a statistical difference was found between mean benzoylecgonine concentrations in vitreous 0.989 mg/L (SD 1.597 mg/L) and blood 1.941 mg/L (SD 2.912 mg/L) ($p = 0.0004$). Regression analysis demonstrated that linear relationships were present between concentrations of vitreous and blood cocaine ($r = 0.854$) and benzoylecgonine ($r = 0.763$). However, the correlation coefficients were lower for cocaethylene ($r = 0.433$) and ethanol ($r = 0.343$). There were variations between the concentrations of cocaine and metabolites both in terms of magnitude and also direction of change. Mean concentrations of benzoylecgonine in blood and vitreous were higher in cases where ethanol was absent, 2.593 mg/L (SD 3.195 mg/L) and 1.431 mg/L (SD 2.021 mg/L), compared to when ethanol was present, 1.199 mg/L (SD 2.396 mg/L) and 0.469 mg/L (SD 0.553 mg/L). This study demonstrates that vitreous humor may be used to quantitate cocaine and cocaine metabolites; however, because the concentrations of cocaethylene in vitreous humor and blood were not well correlated, vitreous humor may not be a reliable specimen for measuring cocaine and cocaine metabolite concentrations.

Marks P.
Med Leg J. 1996;64 (Pt 4):186-93.
Comment in:
Med Leg J. 1997;65 (Pt 3):151-2.
Blood alcohol level: the law and the medicine.
Marraccini JV, Carroll T, Grant S, Halleran S, Benz JA.
Office of the Medical Examiner, Palm Beach County, FL.
J Forensic Sci. 1990 Nov;35(6):1360-6.
Differences between multisite postmortem ethanol concentrations as related to agonal events.

In a study of postmortem ethanol concentrations, blood was withdrawn from the right atrium, ascending aorta, and inferior vena cava. These samples, vitreous humor, and gastric fluid were analyzed in 307 autopsies, where a minimum blood ethanol concentration of 0.05% weight/volume (w/v) was present. Premortem, agonal, and postmortem events were reviewed in an attempt to account for differences in blood ethanol concentrations between sites. The agonal aspiration of vomitus having at least 0.80% w/v ethanol appears to be associated with an increase in aortic ethanol concentrations. We conclude that valid interpretation of postmortem ethanol concentrations must take into consideration the possible entry of ethanol into the pulmonary venous circulation via the respiratory system.

McNeil AR, Gardner A, Stables S.
Department of Molecular Medicine, University of Aukland, New Zealand.
Alanmcn@ahsl.co.nz
Clin Chem.1999 Jan;45(1):135-6.
Simple method for improving the precision of electrolyte measurements in vitreous humor.
Text Available @ *http://www.clinchem.org/cgi/content/full/45/1/135*

Penttila A, Karhunen PJ, Pikkarainen J.
Department of Forensic Medicine, University of Helsinki, Finland.
Forensic Sci Int.1990 Jan;44(1):43-8.
Alcohol screening with the Alcoscan test strip in forensic praxis.

> The Alcoscan test strip was applied as an assay for the screening of alcohol in vitreous humor and urine samples in autopsy cases and in saliva from drunken drivers. The method gives instant and reliable semi-quantitative information on the presence of alcohol and is valuable when considering the necessity of chemical sampling especially during autopsy.

Pleuckhahn VD, Ballard B.
Forensic Sci.1967 Oct;12(4):463-70.
Diffusion of stomach alcohol and heart blood alcohol concentration at autopsy.

Pounder DJ, Kuroda N.
Department of Forensic Medicine, University of Dundee, Scotland.
Forensic Sci Int.1994 Mar 25;65(2):73-80.
Comment in:
> **Forensic Sci Int**. 1995 May 22;73(2):155; author reply 159-60.
> **Forensic Sci Int**. 1995 May 22;73(2):157-8; author reply 159-60.
Vitreous alcohol is of limited value in predicting blood alcohol.

> Vitreous humour alcohol concentration (VHAC) and blood alcohol concentration (BAC) measured by gas chromatography were available from 345 medico-legal autopsies. Simple linear regression with BAC as outcome variable and VHAC as predictor variable (range 1-705 mg%) gave the regression equation $BAC = 3.03 + 0.852\ VHAC$ with 95% prediction interval $+/- 0.019$ square root of $[7157272 + (VHAC - 189.7)2]$ and 99% prediction interval $+/- 0.025$ square root of $[7157272 + (VHAC - 189.7)2]$. The residual standard deviation of VHAC was 26 mg%, the standard error of the slope 0.0098 and the 95% confidence interval for the slope 0.833-0.871. In practice a BAC of 80 mg% is predicted with 95% certainty by a VHAC of 150 mg% and similarly a BAC of 150 mg% by a VHAC of 232 mg%. The prediction interval is too wide to be of real practical use. Previous authors have provided various formulae, including a simple conversion factor, to predict BAC from VHAC without taking into account the uncertainty of the prediction for an individual subject. A re-analysis of the raw data from previous publications gave in most instances regression equations significantly different from the authors' own.

Riggs JE, Schochet SS Jr, Frost JL.
Department of Neurology, West Virginia University School of Medicine,
Morgantown, USA.
Mil Med.1998 Oct;163(10):722-4.
Ethanol level differential between postmortem blood and subdural hematoma.

Alcohol use is a major risk factor for accidental injury and death. However, when death occurs several hours after injury, ethanol in the blood may be absent or low. Ethanol in sequestered hematomas has been used to retrospectively implicate alcohol as a contributing factor at the time of injury. A 69-year-old man died from a large acute subdural hematoma. He had been seen in a hospital emergency department 8 to 12 hours before his death for treatment of two lacerations (one on the head) that occurred during a fall. Postmortem blood ethanol was 0.07%, and subdural hematoma ethanol was 0.04%. This ethanol level differential between the postmortem blood and the subdural hematoma indicates that this man had consumed alcohol after being released from the hospital.

Semenov VA, Shaev AI.
Sud Med Ekspert.1967 Jan-Mar;10(1):8-10.
[On the possibility of photometric determination of alcohol in the blood and urine of cadavers with the use of photoelectrocolorimetry]
[Article in Russian]

Shepherd RT.
Med Leg J.1997;65 (Pt 3):105-6.
Postmortem toxicology.

Sylvester PA, Wong NA, Warren BF, Ranson DL.
University Department of Surgery, University of Bristol, Bristol Royal
Infirmary, UK.
J Clin Pathol.1998 Mar;51(3):250-2.
Unacceptably high site variability in postmortem blood alcohol analysis.

Blood alcohol concentration is a frequently requested test in forensic pathology. The variability of this value was studied by measuring the blood alcohol concentration from six sites in nine subjects at necropsy in whom alcohol was the implicated cause of death. There were small consistent differences in the blood alcohol concentrations between the sites in the nine subjects ($p < 0.04$). Calculation of the mean blood:vitreous humour alcohol concentration ratio (B:V ratio) showed that vitreous humour alcohol concentration most closely reflected the concentration at the femoral vein (B:V ratio = 0.94, r = 0.98), which is considered the optimal site for blood alcohol measurement. The correlation of left heart blood with femoral blood was lower compared with the other sites. There is a potential for an unacceptably large variation in the postmortem measurement of blood alcohol within each subject.

Trojanowska M.
Acta Pol Pharm.1967;24(3):331-4.
[The formation of endogenous ethanol in the blood and urine of cadavers]
[Article in Polish]

Van den Oever R.
Arch Belg Med Soc. 1977 Mar;35(3):181-91.
[Postmortem alcohol concentration in blood and vitreous humor]
[Article in Dutch]

Winek CL, Esposito FM.
Leg Med. 1985;:34-61.
Blood alcohol concentrations: factors affecting predictions.

As a result of extensive alcohol research conducted on both humans and animals, it is possible to predict a BAC, given pertinent data. In addition, it is possible to estimate from a given BAC the quantity of alcohol consumed. Caution must be used in these predictions, for certain factors will affect the final estimation. Absorption of alcohol is influenced by gastrointestinal contents and motility, and also the composition and quantity of the alcoholic beverage. The vascularity of tissues influences the distribution of alcohol, and their water content will determine the amount of alcohol present after equilibrium. Elimination of alcohol begins immediately after absorption. The elimination rate varies for individuals but falls between .015 percent to .020 percent per hour, with an average of .018 percent per hour. In addition to these factors, a BAC will depend on the subject's weight, percentage of alcohol in the beverage, and the rate of drinking. The principal effect of alcohol in the body is on the central nervous system. Its depressant effect consists of impairment to sensory, motor and learned functions. When combined with some other drugs, a more intoxicated state occurs. Although tolerance to alcohol at low blood concentrations is possible, the tolerance most noted is a learned tolerance among chronic drinkers. Contamination of antemortem blood samples collected for alcohol analysis is minimal when swabbing with an ethanolic antiseptic is performed with routine clinical technique; sloppy swabbing has been shown to increase the BAC determination significantly. The alcoholic content of blood used for transfusion does not contribute significantly to the BAC of the recipient, since extensive dilution occurs; nor does the alcohol present in injectable medication contribute significantly. Although many factors may alter the concentration of alcohol present in autopsy specimens, postmortem synthesis of alcohol receives the most attention. The microorganisms that cause postmortem ethanol production can be inhibited by adding a preservative to the samples and storing them under refrigeration. Should putrefaction be present, it is recommended that, in addition to blood, several different specimens be collected and analyzed for the presence of alcohol. Antemortem blood samples containing ethanol, collected using sterile tubes and techniques, may be analyzed up to 14 days later with reasonable certainty that the ethanol level reflects that which was present at the time of collection.

Winek CL, Murphy KL, Winek TA.
Forensic Sci Int. 1984 Aug;25(4):277-81.
The unreliability of using a urine ethanol concentration to predict a blood ethanol concentration.

Of approximately 5,000 forensic cases with a positive ethanol result, over 1,000 were available in which both blood and urine were present for comparison of ethanol content. Data were examined for calculation of the urine to blood ethanol concentration ratio, with the intent of evaluating the validity of predicting a blood ethanol level given a urine ethanol level. The overall urine to blood ethanol concentration ratio was 1.57:1 with a range of 0.7 to 21.0:1. The extremely wide range of values implies that a large degree of error would be introduced if a mean ratio was used when predicting a blood ethanol level from a urine ethanol level.

Additional References

1. Jones AW, Pounder DJ. Measuring blood alcohol concentration for clinical and forensic purposes. In: Karch S, ed. *Handbook of drug abuse*. Boca Raton, Florida: CRC Press, 1997.

2. Garriott JC. Analysis of alcohol in postmortem specimens. In: Garriott JC, ed. *Medicolegal aspects of alcohol determination in biological specimens*. Littleton, Mass: PSG Publishing, 1988.

3. Corry JE. Possible sources of ethanol ante- and post-mortem: its relationship to the biochemistry and microbiology of decomposition. *J Appl Bacteriol* 1978;44:1-56.

4. Nanikawa R, Moriya F, Hashimoto Y. Experimental studies on the mechanism of ethanol formation in corpses. *Z Rechtsmed* 1988;101:21-6.

    Various in vitro experiments were performed for the purpose of clarifying the mechanism of ethanol production in corpses. Whereas a negligible quantity of ethanol was produced in the blood alone, which was left at room temperature, the quantity of ethanol was slightly increased by addition of glucose to the blood. When saprogens were further added, the quantity was markedly increased. Various materials were added to blood-liver homogenates as specimens, and the mixtures were stored in an incubator at 37 degrees C. As a result of the addition of an antibiotic to the mixture every day, there was hardly any production of ethanol. When alcohol dehydrogenase (ADH) and reduced nicotinamide adenine dinucleotide (NADH) were added, ethanol production was slightly increased. When acetaldehyde was added first, ethanol production was inhibited the next day, but on and after day 2, the quantity of ethanol was more than that in the control material. When pyruvic acid was added first, the results were similar to the above. Pyrazole, cyanamide, and disulfiram completely inhibited the production of ethanol. Ethanol production in corpses is believed to take place through a pathway opposite to that of ethanol metabolism in the living body, under the influence of ADH, ALDH, etc., in saprogens using carbohydrates as substrates.

5. Zumwalt RE, Bost RO, Sunshine I. Evaluation of ethanol concentrations in decomposed bodies. *J Forensic Sci* 1982;27:549-54.

6. Canfield DV, Kupiec T, Huffine E. Postmortem alcohol production in fatal aircraft accidents. *J Forens Sci* 1993;38:914-7.

7. Pounder DJ, Kuroda N. Vitreous alcohol is of limited value in predicting blood alcohol. *Forens Sci Int* 1994;65:73-80.

8. Kuroda N, Williams K, Pounder DJ. Estimating blood alcohol from urinary alcohol at autopsy. *Am J For Med Path* 1995;16:219-22.

9. Harper DR. A comparative study of the microbiological contamination of postmortem blood and vitreous humour samples taken for ethanol determination. *Forens Sci Int* 1989;43:37-44.

10. Levine B, Smith ML, Smialek JE, Caplan YH. Interpretation of low postmortem concentrations of ethanol. *J Forens Sci* 1993;38:663-7.

11. Mayes R, Levine B, Smith ML, Wagner GN, Froede R. Toxicological findings in the USS Iowa disaster. *J Forens Sci* 1992;37:1352-7.

12. Marraccini JV, Carroll T, Grant S, Halleran S, Benz JA. Differences between multisite postmortem ethanol concentrations as related to agonal events. *J Forens Sci* 1990;35:1360-6.

13. Pounder DJ, Smith DRW. Postmortem diffusion of alcohol from the stomach. *Am J Forens Med Path* 1995;16:89-96.

14. Briglia EJ, Bidanset JH, Dal Cortivo LA. The distribution of ethanol in postmortem blood specimens. J Forens Sci 1992;37:991-8.

15. Knight B. Forensic pathology. London: Edward Arnold, 1991:493-4.

16. Cox DE, Sadler DW, Pounder DJ. Alcohol estimation at necropsy. J Clin Pathol 1997;50:197-201.

AIMS: To gather data on blood alcohol concentrations in a forensic necropsy population and to analyse the information on trends that may predict where alcohol testing is going to prove cost-effective. METHODS: Alcohol assays were performed on blood, urine, and vitreous samples in 1620 consecutive medicolegal necropsy examinations. RESULTS: Alcohol was detected in only 7% of natural deaths from all causes and in four of 40 deaths categorised as unknown/obscure. Alcohol concentrations > or = 350 mg/100 ml were found in nine drug/alcohol abuse deaths (range 362-506 mg/100 ml), five accidental deaths (356-504 mg/100 ml), and one homicide victim (400 mg/100 ml). Those categorised as alcohol abusers were represented in all but one category of death (unknown/obscure deaths in males), showing that many true alcoholics die with their alcoholism rather than of it; 39% of males and 34% of females with histories of alcohol abuse had alcohol present in their blood at necropsy at concentrations > or = 50 mg/100 ml, v only 9% (male) and 6% (female) without such history. CONCLUSIONS: The study highlights the problems of elderly and "hidden" alcoholics and illustrates cases where routine assays would provide additional significant information. Routine alcohol testing is useful in all cases of suspected unnatural death but universal testing of forensic necropsies is not cost-effective.

Website Information

http://www.coheadquarters.com/coFire/cofire4.htm
Death from CO in Fire and Non-Fire Situations: Role of Alcohol (Ethanol, Ethyl Alcohol, ETOH) in Death from CO

http://www.vv.se/traf_sak/t2000/POSTER7.pdf
"Blood alcohol concentrations in an autopsy material in practice of the Institute of Forensic Research in 1990-1999"
Gubala, W.; Piekoszewski, W. (Poland)

http://bmj.bmjjournals.com/cgi/content/full/316/7125/87
BMJ 1998;316:87 (10 January)
"Dead sober or dead drunk? May be hard to determine"

Carbon Monoxide Determination During Autopsy (abstracts provided where available)

Barillo DJ, Rush BF Jr, Goode R, Lin RL, Freda A, Anderson EJ Jr.
Am Surg.1986 Dec;52(12):641-5.
Is ethanol the unknown toxin in smoke inhalation injury?

Of the 12,000 fire-related deaths occurring annually in the United States, it is estimated that 60 to 80 per cent are due to smoke inhalation. Plastic and synthetic materials which have been introduced in home construction and furnishings produce a more toxic smoke when burned. Efforts to identify a "supertoxin" in this smoke have been unsuccessful to date. An alternative approach is to examine why victims are unable to escape, and become exposed to smoke for lethal periods of time. The authors examined the circumstances of death in 39 fire victims (27 adults, 12 children) over a 25-month period. Detailed examination of the fire scene, autopsy studies, and toxicologic analysis were carried out. Position of the victim, and escape efforts were noted. Carbon monoxide was elevated in all victims, with "lethal" levels (=greater than 50%) in 21/39 victims. Cyanide was detected in 24/29 victims, but none had lethal (3 mg/L) levels present. Ethanol was detected in 21/26 adults (80%) and 0/12 children (0%). 18/26 adult victims had ethanol levels above the statutory level of intoxication (10 mg%). Victims found in bed (no escape attempt) had a mean blood ethanol level of 268 mg%, compared with a mean level of 88 mg% in those victims found near an exit (P = .006). Ethanol intoxication significantly impairs the ability to escape from fire and smoke and is a contributory factor in smoke-related mortality.

Blackmore DJ.
Analyst.1970 May;95(130):439-58.
The determination of carbon monoxide in blood and tissue.
Chen KC, Lee EW, McGrath JJ.
J Appl Toxicol.1984 Jun;4(3):145-9.
Effect of intermittent carbon monoxide inhalation on erythropoiesis and organ weights in rats.

Sprague-Dawley rats were exposed to 450 ppm carbon monoxide (CO) for 6 h per day, 5 days per week for 33 days. The effect of CO on reticulocyte count, hematocrit, hemoglobin concentration, body weight and selected organ weights was measured. Exposure to CO caused a three-fold increase in the youngest reticulocyte population, concomitant with an increase in the total reticulocyte count. Despite continued CO exposure, reticulocyte number and distribution returned to normal by day 9, suggesting that reticulocyte response of the organism to CO had changed. Both hematocrit and hemoglobin concentrations began to increase 16 days after CO exposure and remained at the increased level for the duration of the exposure period. There were no changes in kidney, liver and adrenal weights throughout the course of study. However, spleen weight was increased after 5 days of CO exposure. Left and right ventricular organ weight ratios increased equally at the same time during the study. These results indicate that the increase in the young reticulocyte population and the subsequent increase in total reticulocyte count are the earliest erythropoietic responses to intermittent CO exposure and that CO-induced polycythemia is associated with cardiac hypertrophy in rats.

Christensen AM, Icove DJ.
U.S. Tennessee Valley Authority Police, 400 W. Summit Hill Dr., WT-3D,
Knoxville, TN, USA
J Forensic Sci.2004 Jan;49(1):104-7.
The application of NIST's Fire Dynamics Simulator to the investigation of carbon monoxide exposure in the deaths of three Pittsburgh fire fighters.

A case is reported in which computer fire modeling was used to reevaluate a fire that killed three fire fighters. NIST's Fire Dynamics Simulator (FDS) was employed to model the fire in order to estimate the concentration of carbon monoxide present in the dwelling, which was the immediate cause of death of two of the fire fighters, who appear to have removed their face pieces in order to share available air. This estimate, along with an assumed respiration volume and known blood carboxyhemoglobin, was plugged into a standard equation to estimate the time of exposure. The model indicated that 27 min into the fire, the carbon monoxide concentration had already reached approximately 3600 ppm. At this concentration, and a respiration of 70 L/min, an estimated 3 to 8 min of exposure would have been required to accumulate the concentrations of carboxyhemoglobin (49, 44, and 10%) measured on the fire fighters at autopsy.

Freireich AW, Landau D.
J Forensic Sci.1971 Jan;16(1):112-9.
Carbon monoxide determination in postmortem clotted blood.

Grabowska T, Sybirska H, Malinski M.
Katedry Medycyny Sadowej Slaskiej AM w Katowicach
Arch Med Sadowej Kryminol.2003 Jan-Mar;53(1):9-17.
[Attempt to estimate risks of fatal poisoning on the basis of HCN and HbCO concentrations in blood of fire victims]
[Article in Polish]

Using the results of HCN and HbCO concentrations in the blood of 174 deceased found in different burning spaces and 35 people with symptoms of poisoning evacuated from the scene of a fire and then admitted to hospital. The correlation between blood concentration of both these xenobiotics and death or chance of survival in a fire was estimated by statistical analysis. An attempt was made to define a value of so-called "cut-off" points for HbCO and HCN by independence test chi 2 with Yates's correction. Point and interval estimations (95% Cornfield's confidence interval) were used for the odds ratio (OR). The research showed that there was a strict statistical correlation between the chance of survival and death risks dependent on blood concentrations of HCN and HbCO in all the groups examined.

Hirsch CS, Adelson L.
JAMA.1969 Dec 22;210(12):2279-80.
Absence of carboxyhemoglobin in flash fire victims.

Iffland R, Sticht G.
Arch Toxikol.1972;29(4):325-30.
[Gas chromatographic method for determination of carbon monoxide in blood]
[Article in German]

Jones JS, Lagasse J, Zimmerman G.
Emergency Medicine Residency Program, Butterworth Hospital, Grand Rapids, MI.
Am J Emerg Med.1994 Jul;12(4):448-51.
Computed tomographic findings after acute carbon monoxide poisoning.

Selective necrosis and degeneration of the globus pallidus are characteristic autopsy findings in patients with severe carbon monoxide (CO) poisoning. The objective of this study was to show that computed tomography (CT) may demonstrate these morphological changes in the brain during life, and provide a clue to prognosis. The authors reviewed the medical records of 19 consecutive patients with acute CO poisoning who underwent CT examination during hospitalization. Abnormal CT findings were found in 10 of the 19 patients (53%). The most common abnormal findings were low-density areas in the basal ganglia. These lesions were found in 7 of the 10 cases, and varied from small (limited to the globus pallidus) to large (extending to the internal capsule). Of the 10 patients with abnormal CT scans, 9 survived to hospital discharge but all had some degree of functional neurological impairment. Eighty-nine percent (8 of 9) of the patients with normal CT scans were discharged neurologically intact. Awareness of the potential for basal ganglia lesions in CO poisoning should lead to more accurate CT interpretation and may have significant prognostic implications

Kojima T, Nishiyama Y, Yashiki M, Une I.
Forensic Sci Int.1982 May-Jun;19(3):243-8.
Postmortem formation of carbon monoxide.

Since carbon monoxide (CO) production after death was suggested in a drowned body, CO and carboxyhemoglobin (HbCO) levels in blood and body cavity fluids of cadavers which were not exposed to fire and CO hve been analyzed. CO released from the tissues was determined by gas chromatography and gas chromatography-mass spectrometry, and the total concentration of hemoglobin (Hb) was measured as cyanmethemoglobin (CNmHb). The HbCO level was calculated by the ratio of CO content and CO-binding capacity. CO levels (ml/100 g at STP) of the seven cases in which blood and body cavity fluids could be collected ranged from 0.13 to 0.87 in blood and 0.02 to 0.80 in body cavity fluids. HbCO levels in blood and body cavity fluids were from 0.3 to 6.0% and from 2.3 to 44.1%, respectively. In a typical case showing postmortem formation of CO, the CO levels in body cavity fluids were higher than that in blood. It is suggested that CO in a putrefied body is due to CO in blood prior to death and the CO formed by the decomposition of Hb, myoglobin and other substances during putrefaction. The significance of HbCO levels in body cavity fluids of cases with marked postmortem decomposition seems difficult to interpret without the value of HbCO in blood.

Kuller LH, Radford EP, Swift D, Perper JA, Fisher R.
Arch Environ Health.1975 Oct;30(10):477-82.
Carbon monoxide and heart attacks.

A study of the relationship between carbon monoxide exposure and heart attacks was conducted in Baltimore. There was no evidence of clustering of either myocardial infarction or sudden ASHD on a specific day, nor was there correlation between the number of cases per day and ambient CO levels. Postmortem HbCO levels were slightly higher in ASHD sudden deaths than in sudden deaths due to other causes. Any differences were probably primarily due to cigarette smoking. Cigarette smokers who died suddenly due to ASHD had substantially higher postmortem HbCO levels than nonsmokers. Practically all of the elevated HbCO levels could be related to cigarette smoking or specific environmental exposure. There were no differences between HbCO levels in ASHD sudden death patients and in living controls. There was also no relationship between cardiac pathologic findings and postmortem HbCO levels among patients dying suddenly of ASH.

Kunsman GW, Presses CL, Rodriguez P.
Forensic Toxicology Laboratory, Bexar County Medical Examiner's Office, San
Antonio, Texas 78229, USA.
J Anal Toxicol.2000 Oct;24(7):572-8.
Carbon monoxide stability in stored postmortem blood samples.

Carbon monoxide (CO) poisoning remains a common cause of both suicidal and accidental deaths in the United States. As a consequence, determination of the percent carboxyhemoglobin (%COHb) level in postmortem blood is a common analysis performed in toxicology laboratories. The blood specimens analyzed are generally preserved with either EDTA or sodium fluoride. Potentially problematic scenarios that may arise in conjunction with CO analysis are a first analysis or a reanalysis requested months or years after the initial toxicology testing is completed; both raise the issue of the stability of carboxyhemoglobin in stored postmortem blood specimens. A study was conducted at the Bexar County Medical Examiner's Office to evaluate the stability of CO in blood samples collected in red-, gray-, and purple-top tubes by comparing results obtained at the time of the autopsy and after two years of storage at 3 degrees C using either an IL 282 or 682 CO-Oximeter. The results from this study suggest that carboxyhemoglobin is stable in blood specimens collected in vacutainer tubes, with or without preservative, and stored refrigerated for up to two years.

Malik MO.
J Forensic Sci Soc.1971 Jan;11(1):21-8.
Problems in the diagnosis of the causes of death in burned bodies.

Mallach HJ, Mittmeyer HJ.
Beitr Gerichtl Med.1979;37:393-9.
[Effect of exposure, alcohol and body disposition on the carbon monoxideconcentration of fatal poisonings]
[Article in German]

Mayes RW.
RAF Institute of Pathology and Tropical Medicine, Halton Aylesbury.
J Clin Pathol.1993 Nov;46(11):982-8.
ACP Broadsheet No 142: November 1993. Measurement of carbon monoxide and cyanide in blood.

Miyazaki T, Kojima T, Yashiki M, Chikasue F, Iwasaki Y.
Department of Legal Medicine, Hiroshima University School of Medicine, Japan.
Int J Legal Med.1992;105(2):65-8.
Interpretation of COHb concentrations in the left and right heart blood of cadavers.

Carbon monoxide hemoglobin (COHb) concentrations in left and right heart blood samples from cadavers both exposed and not exposed to fire or CO gas were analyzed by the gas chromatographic method. The COHb concentration ratio between samples of left and right heart blood (L/R ratio) does not appear to be useful for establishing whether death has occurred before or after exposure to fire with the exception of cases where no soot can be detected in the airways by the naked eye and the COHb concentration in the blood sample is within the level considered normal for tobacco smokers.

Morinaga M, Kashimura S, Hara K, Hieda Y, Kageura M.
Department of Forensic Medicine, Fukuoka University School of Medicine, Japan.
Int J Legal Med.1996;109(2):75-9.
The utility of volatile hydrocarbon analysis in cases of carbon monoxide poisoning.

A new approach to investigate the circumstances relating to carbon monoxide intoxication by analysing volatile hydrocarbons in the blood of cadavers is reported. Headspace gas chromatography/mass spectrometry was used to demonstrate the hydrocarbons. The results can be characterized into four categories depending on the compounds detected. In construction fire cases where no accelerants were found at the scene benzene, toluene and styrene were detected in the blood. In cases where gasoline was found in the fire debris surrounding the victim, high levels of benzene, toluene, ethylbenzene, xylene isomers, n-hexane and n-heptane were detected in the blood. In cases where kerosene was found in the fire debris around the victim, benzene, toluene, ethylbenzene, xylene isomers, C9-aromatics(n-propylbenzene, trimethyl-benzene isomers), n-octane, n-nonane and n-decane were detected in the blood. In cases where the victim was found inside a gasoline-fuelled automobile filled with exhaust gas, benzene, toluene, ethylbenzene, xylene isomers, C9-aromatics were found, but no aliphatic hydrocarbons such as components of petroleum. The analyses of the combustion gases of inflammable materials, exhaust gas, gasoline vapours and kerosene vapours were also performed to evaluate the results of the blood analyses. Consequently, some compounds are proposed as indicators to discriminate between inhaled gases i.e. styrene in common combustion gas, n-hexane and n-heptane as well as benzene, toluene and C9-aromatics in gasoline cases, n-nonane and n-octane as well as benzene, toluene and C9-aromatics in kerosene cases, and benzene, toluene, C9-aromatics but no aliphatic hydrocarbons in exhaust gas cases.

Nakatome M, Matoba R, Ogura Y, Tun Z, Iwasa M, Maeno Y, Koyama H, Nakamura Y, Inoue H.
Department of Legal Medicine, Course of Social Medicine, Osaka University Graduate School of Medicine, Suita City, Japan. nakatome@legal.med.osaka-u.ac.jp
Int J Legal Med.2002 Feb;116(1):17-21.
Detection of cardiomyocyte apoptosis in forensic autopsy cases.

The purpose of the present study was to determine reliable parameters for the detection of apoptotic cells for use as a diagnostic marker during the early stage of acute myocardial infarction (AMI) in forensic autopsy cases. Myocardial tissues taken from forensic autopsy cases were examined by immunohistochemical and molecular-biological methods using the terminal deoxynucleotidyl transferase-mediated dUTP biotin nick end-labelling (TUNEL) and the DNA laddering methods. In cases of AMI with a time period between 2 h from onset to death and 20 h post-mortem time, the nuclei of cardiomyocytes were stained positive with the TUNEL method and DNA fragmentation of myocardial cells was detected by agarose gel electrophoresis. Similar findings were obtained in cases of carbon monoxide (CO) intoxication. However, no apoptotic cells were found in other cases such as methamphetamine (MAP) intoxication, tetrodotoxin intoxication, alcohol intoxication, asphyxia, head injury, heart injury or myocarditis. These findings suggested that it would be possible to apply TUNEL-positive cells as a diagnostic marker during the early stages of AMI.

Perrigo BJ, Joynt BP.
Central Forensic Laboratory, Royal Canadian Mounted Police, Ottawa.
J Anal Toxicol.1989 Jan-Feb;13(1):37-46.
Evaluation of current derivative spectrophotometric methodology for the determination of percent carboxyhemoglobin saturation in postmortem blood samples.

Carbon monoxide intoxication continues to be a commonly encountered cause of death in most areas of Canada. The forensic nature of the samples in these cases presents special problems that are not normally encountered in clinical determinations. A study was undertaken to assess various methods of determining the percent carboxyhemoglobin saturation in blood, more specifically, those using derivative spectrophotometric measurements in the Soret region of the UV spectrum. At the same time, other studies were carried out: the effects of storage time on the carboxyhemoglobin levels; evaluation of sample containers; comparison of percent carboxyhemoglobin saturation in blood samples taken ante-mortem and post-mortem. Blood for the study was obtained from laboratory animals that were exposed to carbon monoxide before death.

Oritani S, Zhu BL, Ishida K, Shimotouge K, Quan L, Fujita MQ, Maeda H.
Department of Legal Medicine, Osaka City University Medical School, Asahi-machi
1-4-3, Abeno, 545-8585, Osaka, Japan
Forensic Sci Int.2000 Sep 11;113(1-3):375-9.
Automated determination of carboxyhemoglobin contents in autopsy materials usinghead-space gas chromatography/mass spectrometry.

To establish a method for the routine analysis of carboxyhemoglobin (COHb) in autopsy materials including those which have undergone postmortem changes, e.g. thermo coagulation, putrifaction and contamination, an automated head-space gas chromatography/mass spectrometry (GC/MS) analysis was utilized. The procedure consisted of preparation of the sample in a vial and a carbon monoxide (CO) saturated sample, for estimation of hemoglobin content, in another vial, the addition of n-octanol, potassium ferricyanide and an internal standard (t-butanol), GC separation and determination of CO using a GC/MS system equipped with an automated head-space gas sampler. The method was practical not only with the blood and bone marrow aspirates to confirm the findings on the CO-oximeter system, but also with the thermo-coagulated and putrified blood.

Quan L, Zhu BL, Fujita MQ, Maeda H.
Department of Legal Medicine, Osaka City University Medical School, Asahi-machi
1-4-3, Abeno, 545-8585 Osaka, Japan. legalmed@med.osaka-cu.ac.jp
Leg Med (Tokyo).2003 Mar;5 Suppl 1:S335-7.
Ultrasonographic densitometry of the lungs at autopsy: a preliminary investigation for possible application in forensic pathology.

The aim of the present study was to examine the possible application of ultrasonographic (US) densitometry of the lungs to quantitative evaluation of pulmonary edema at autopsy (n=85). A diagnostic ultrasound device LOGIQ alpha200 (GE Yokogawa Medical Systems) equipped with an LH probe (linear, 7.5 MHz) was used and each lobe of the lungs was scanned on the anterior and posterior surfaces after resection. The US density showed a correlation between the left and right lobes, and also between the anterior and posterior surface scans of each lobe. Although there was a correlation between the US density and combined lung weight in total cases, the density ranged very widely when lung weight was below about 1300 g, depending on the cause of death. The density was high in drowning, asphyxia, poisoning and delayed traumatic death, whereas it was usually low in fire death mainly due to burns, hemorrhagic shock and head injury. In the other causes of death, a considerable case-to-case difference was observed independent of the lung weight. These findings suggested a possible contribution of pulmonary edema to high US density, possibly depending on the survival time and irrespective of the blood contents (congestion or postmortem hypostasis).

Quan L, Zhu BL, Oritani S, Ishida K, Fujita MQ, Maeda H.
Department of Legal Medicine, Osaka City University Medical School, Japan.
Int J Legal Med.2001;114(6):310-5.
Intranuclear ubiquitin immunoreactivity in the pigmented neurons of the substantia nigra in fire fatalities.

To evaluate the significance of immunohistochemical staining of ubiquitin (heat shock protein) in the midbrain for medico-legal investigation of death in fires, we examined forensic autopsy cases of fire fatalities (n = 35) in comparison with controls (n = 27; brain stem injury, acute myocardial infarction and carbon monoxide poisoning other than fire fatality). There were two intranuclear staining patterns in the nuclei of pigmented substantia nigra neurons: a type of inclusion (possible Marinesco bodies) and a diffuse staining. Percentage of nuclear ubiquitin positivity (Ub-positive %) in fire fatalities (2.7-44.7%; mean, 18.5%) was significantly higher than in brain stem injury (n = 9; 0-10.4%; mean, 4.5%) and myocardial infarction (n = 14; 1.5-14.6%; mean, 6.9%), independently of blood carboxyhemoglobin (COHb) levels. Age-dependent increase in Ub-positive % was observed in lower COHb (< 60%) cases. The intranuclear diffuse ubiquitin staining was not observed in cases of high blood cyanide level (> 1.0 microg/ml). These observations showed that intranuclear ubiquitin immunoreactivity of the pigmented substantia nigra neurons in the midbrain was induced by severe stress in fires.

Reys LL, Santos JC.
Institute of Legal Medicine, University School of Medicine, Lisbon, Portugal.
Am J Forensic Med Pathol.1992 Mar;13(1):33-6.
Importance of information in forensic toxicology.

Information in forensic toxicology plays a very important role. The forensic pathologist usually seeks toxicologic analyses on basis of the information available at the time of the medicolegal autopsy. Such information may be obtained from different sources: hospitals, authorities, relatives, friends, or neighbors of the deceased and, obviously, macroscopic findings at the time of the autopsy. In order to evaluate the relative importance of these different sources of information, the authors have studied, retrospectively, results of 580 postmortem examinations performed at the Institute of Legal Medicine of Lisbon, wherein toxicologic analyses had been requested. These cases pertain to the years 1987 and 1988, but do not include alcohol determination in the blood in cases of traffic accidents. In 274 (47.4%) of the 580 cases, there were positive findings while in the remaining 306 (52.6%) findings were negative. In cases with positive findings, circumstances and factors, which may have influenced the pathologist's decision to request toxicologic analysis, are discussed. In more than half the cases, hospital information was the decisive factor, while in approximately 25% of the cases, autopsy findings were the justification. In contrast, it is worth mentioning that in approximately 45% of the cases with analytical negative results, requests were made, in cases of blank autopsies, for toxicologic analyses in order to exclude the possibility of poisoning. It is interesting to note that in the same proportion requests were justified on grounds of hospital information. Some of the factors that may explain this apparent discrepancy are discussed. Finally, the relevance of background information is emphasized at the level of the interpretation of analytical results, whether positive or negative.

Schmidt P, Musshoff F, Dettmeyer R, Madea B.
Institut fur Bechtsmedizin, Universitat Bonn.
Arch Kriminol.2001 Jul-Aug;208(1-2):10-23.
[Unusual carbon monoxide poisoning]
[Article in German]

Despite of indicative death scenes or characteristic findings of the external examination, about 40% of the accidental fatal intoxications due to carbon monoxide are not recognized before the performance of the autopsy. Six cases are reported which illustrate possible reasons for the delayed establishment of the diagnosis: unusual circumstances of the intoxication or sources of carbon monoxide, only subtle degree or lack of external signs of the intoxication or a competing cause of death at autopsy.--Cases 1 and 2: 53, respectively 54-year-old couple, found dead in a caravan, extreme putrefaction of the bodies, spectrophotometric detection of the fatal carboxyhaemoglobin level in oedema fluid of the scalp.--Case 3: 23-year-old lorry driver, found dead in the tightly closed cab of his lorry, operation of a source of electricity with "environmentally friendly" fuel, carboxyhaemoglobin level 83%.--Case 4: 19-year-old man, found dead in the flat of friends, removal of the CO-source before alerting the police forces, lack of the bright pink coloration of livor mortis, haemopericardium due to atrial rupture at postmortem examination, carboxyhaemoglobin level 65%.--Case 5: 27-year-old man, found dead in his flat, advanced decomposition of the body, residues of a charcoal fire in a metal bucket in the sink, carboxyhaemoglobin level 80%.--Case 6: 42-year-old woman, lying dead in the garage beside her car, engine switched-off, ignition key next to the body on the floor under the car, carboxyhaemoglobin level 46%.

Shinomiya T, Shinomiya K.
Acta Med Leg Soc (Liege).1989;39(1):131-43.
[The variation in carbon monoxide release in the blood stain and in visceral tissues]
[Article in French]

Teige B, Lundevall J, Fleischer E.
Z Rechtsmed.1977 Jul 5;80(1):17-21.
Carboxyhemoglobin concentrations in fire victims and in cases of fatal carbon monoxide poisoning.

The study comprises an eleven-year autopsy material of 141 cases from the Institute of Forensic Medicine, Oslo. The fatal level of carboxyhemoglobin concentration is calculate from cases of pure carbon monoxide poisoning. Carboxyhemoglobin concentrations below this level are found in approximately thirty percent of the fire victims. Alcohol intoxication, present in many fire victims, is not related to low corboxyhemoglobin concentrations.

von Meyer L, Drasch G, Kauert G.
Z Rechtsmed.1979;84(1):69-73.
[Significance of hydrocyanic acid formation during fires]
[Article in German]

Cyanide concentrations of blood samples from fire victims autopsied in the Institute of Legal Medicine, Munich, have been determined. In 25% of 48 analyzed cases cyanide concentrations from 0.52 microgram to 6.24 microgram Cyanide/ml blood have been detected. These results are compared to former studies and the higher mean level in our collective is emphasized. The importance of hydrocyanid acid in the toxicity of fire gases is evidently greater, than assumed. Hydrocyanic acid may be produced from nitrogen continaing polymers during combustion. The quote of these polymers in clothing, furniture, and also in equipment of cars is increasing. Therefore, it is necessary to take more notice of the formation of hydrocyanic acid during combustion, even though carbon monoxide is in general the main toxic agent in fire gases.

Winek CL, Prex DM.
Forensic Sci Int.1981 Sep-Oct;18(2):181-7.
A comparative study of analytical methods to determine postmortem changes in carbon monoxide concentration.

Twenty-one autopsy blood samples were analyzed using spectrophotometric and gas chromatographic procedures after storage for 30 and 150 days. When carboxyhemoglobin was measured spectrophotometrically at the absorbance ratio of 540 nm/555 nm, the observed average percent losses were 8 +/- 9% and 35 +/- 27% after 30 and 150 days of storage, respectively. When measured at the absorbance ratio of 540 nm/579 nm, the average percent losses of carboxyhemoglobin were 7 +/- 8% and 34 +/- 25% after 30 and 150 days, respectively. Wavelength shifts and distorted spectral scans were observed at 150 days. When carbon monoxide was determined by gas chromatographic methods based on combining capacity, the average percent loss was 15 +/- 24% and 37 +/- 36% after 30 days and 150 days, respectively. The average percent loss of calculated CO based on hemoglobin concentration after 30 days was 31 +/- 14% and at 150 days, 40 +/- 24%. The average percent loss of calculated CO based on iron content was 23 +/- 13% and 37 +/- 23% after 30 and 150 days, respectively.

Wirthwein DP, Pless JE.
Division of Forensic Pathology, Indiana University, Indianapolis 46202-5120, USA.
Am J Forensic Med Pathol.1996 Jun;17(2):117-23.
Carboxyhemoglobin levels in a series of automobile fires. Death due to crash or fire?

The determination of death by trauma versus fire can be of major consideration, especially in civil product liability litigation. Blood carboxyhemoglobin levels can be instrumental in that differentiation. Twenty-eight fatalities involving fire in automobiles were reviewed. All subjects displayed some degree of body burn, and in 25 severe charring and/or incineration was present at autopsy. In only one case was there a history of explosion or flash fire. Carboxyhemoglobin levels varied from 92% to values of < 10%. In seven cases no collision occurred. In six of these subjects COHb values were > or = 47%. In all 16 cases with carboxyhemoglobin levels of < or = 10% a collision occurred. In 12 of 16 of these subjects, blunt force injury sufficient to cause death was discovered. Data presented in this article indicate that a carboxyhemoglobin level of > 30% strongly suggests inhalation of combustion products as the cause of death. In contrast, a level of < 20% should prompt a search for other causes.

Wu SC, Levine B, Goodin JC, Caplan YH, Smith ML.
Office of the Chief Medical Examiner, State of Maryland.
J Anal Toxicol.1992 Jan-Feb;16(1):42-4.
Analysis of spleen specimens for carbon monoxide.

Crucial to the investigation of aircraft fatalities is the analysis of biological specimens for carbon monoxide (CO). In many cases, blood specimens are unavailable or unsuitable for analysis, and the testing of an alternate specimen for CO becomes necessary. Spleen specimens provide a rich source of red blood cells and hence can be a primary substitute for blood. To verify this, 40 paired blood and spleen specimens were analyzed for CO by using a gas chromatographic method. Ten specimens with a spleen CO saturation level (sat.) of less than 10% were associated with corresponding blood specimens with CO sat. less than 10%. Fifteen of the 18 spleen specimens with CO sat. greater than 29% were associated with blood specimens with greater than 48% sat. Results were inconclusive when the spleen CO sat. was between 10 and 29%. We concluded that spleen CO sat. can reflect blood CO sat. in certain situations, particularly when spleen CO sat. is high.

Appendix A: Death Investigation Systems In The United States And Territories

State/ Territory	Type of System	Title	Deaths Investigated	Contact	Comment
Alabama	Mixed: State Medical Examiner and County Coroners/Medical Examiners	County Coroner, State Medical Examiner	Coroner—All deaths where the deceased died without being attended by a legally qualified physician. State Medical Examiner— • If the person dies by violence or homicide, suicide, accidental, or industrial. • Criminal abortion. • Sudden death, if in apparent good health. • In suspicious circumstances. • When a public health hazard. • If the body is to be cremated.	Emily Ward, MD Alabama Department of Forensic Sciences P.O. Box 240591 Montgomery, AL 36124-0591 (334) 242-3093 Fax: (334) 260-8734	Slight differences in death investigation in Jefferson County (Birmingham) from those in other counties.
Alaska	State Medical Examiner	State Medical Examiner	• Homicide. • Suspicion of criminal means. • Suicide. • With no physician in attendance. • When physician is unable to execute death certificate. • When cause of death cannot be determined.	Franc G. Fallico, MD Acting State Medical Examiner 4500 South Boniface Pkwy. Anchorage, AK 99507 (907) 269-5090 Fax: (907) 334-2216 Email: Franc_Fallico@ HEALTH.STATE.AK.US	

State/ Territory	Type of System	Title	Deaths Investigated	Contact	Comment
American Samoa	Territorial Coroner	No death investigation official.	• Where any dead body is found. • By accidental means. • Allegedly caused by unlawful means.	Malaetasi M. Cogafau Attorney General of American Samoa P.O. Box 7 Pago Pago, AS 96799 (684) 633-4163	Pulenuu (village chief) reports deaths to the Attorney General, or, if necessary, to the Department of Medical Services' local representative, who is authorized to act as Coroner and report findings to the Attorney General.
Arizona	County Medical Examiner	County Medical Examiner	• When not under current care of a physician for a potentially fatal illness. • When the attending physician is unavailable to sign the death certificate. • By violence. • Of a prisoner or occurring in prison. • Occurring suddenly to a person in apparent good health. • By occupational disease or accident. • Where a public health hazard is presented. • Occurring during anesthetic or surgical procedures. • Occurring in a suspicious, unusual, or unnatural manner.	Philip E. Keen, M.D. Medical Examiner Maricopa County 120 South Sixth Avenue Phoenix, AZ 85003 (602) 506-3322 Fax: (602) 506-1546	
Arkansas	Mixed: State Medical Examiner and County Coroners/Medical Examiners	County Coroner, State Medical Examiner	• Death appears to be caused by violence, homicide, suicide, or accident. • Death appears to be the result of the presence of drugs or poisons. • Death appears to be the result of a motor vehicle accident or the body was found in or near a roadway or railroad.	William Q. Sturner, M.D. State Medical Examiner P.O. Box 8500 Little Rock, AR 72215 (501) 227-5936 Fax: (501) 221-1653 Email: William.Sturner@ ASCL.State.AR.US	

State/ Territory	Type of System	Title	Deaths Investigated	Contact	Comment
Arkansas (cont'd)			• Death appears to be the result of a motor vehicle accident and there is no obvious trauma to the body.		
			• Death occurs while the person is in a State mental institution or hospital and there is no previous medical history to explain the death, or while the person is in police custody, a jail, or penal institution.		
			• Death appears to be the result of a fire or explosion.		
			• Death appears to be the result of drowning.		
			• Death of a minor child appears to indicate child abuse.		
			• Death of an infant or minor child without previous medical history.		
			• Human skeletal remains are recovered.		
			• Decomposition of the body prohibits external examination to rule out injury or circumstances of death cannot rule out the commission of a crime.		
			• Manner of death appears to be other than natural.		
			• Death is sudden and unexplained.		
			• Death occurs at a worksite.		
			• Death is due to criminal abortion.		
			• A physician was not in attendance within 36 hours preceding death or, in pre-diagnosed terminal or bedfast cases, within 30 days. This includes persons admitted to an emergency room, unconscious and unresponsive, following cardiopulmonary resuscitation, who die within 24 hours of admission without regaining consciousness or responsiveness.		

State/ Territory	Type of System	Title	Deaths Investigated	Contact	Comment
California	Mixed: County Medical Examiners/ Coroners	County Coroner, County Sheriff-Coroner, County Medical Examiner, Coroner	• Violent, sudden, or unusual. • Unattended. • Deaths wherein the deceased has not been attended by a physician in the 20 days before death. • Self-induced or criminal abortion. • Known or suspected homicide, suicide, or accidental poisoning. • By recent or old injury or accident. • Drowning, fire, hanging, gunshot, stabbing, cutting, exposure, starvation, acute alcoholism, drug addiction, strangulation, aspiration. • Suspected sudden infant death syndrome. • By criminal means. • Associated with known or alleged rape or crime against nature. • By known or suspected contagious disease constituting a public hazard. • By occupational disease or hazard. • Of State mental hospital patient. • Of developmentally disabled patient in State developmental services hospital. • In prison or while under sentence. • Under other suspicious circumstances.	Scotty D. Hill Executive Secretary California State Coroners Association 5925 Maybrook Circle Riverside, CA 92506-4549 (909) 788-2656 Fax: (909) 788-2934 Email: CSCA2000@aol. com	41 counties with Sheriff-Coroners. In addition, many County Coroners also serve as Public Administrator, Public Guardian, and Public Conservator.

State/ Territory	Type of System	Title	Deaths Investigated	Contact	Comment
Colorado	County Coroner	County Coroner	• From external violence. • Unexplained cause. • Under suspicious circumstances. • Suddenly, when in good health. • Where no physician is in attendance, or where the attending physician is unable to certify the cause of death. • From thermal, chemical, or radiation injury. • From criminal abortion. • From disease that may be hazardous or contagious or may constitute a hazard to the public health. • While in custody of law enforcement officials or while incarcerated in a public institution. • From industrial accident.	James L. Kramer, P.A.C. President Colorado Coroner Association 517 Colorado Avenue Pueblo, CO 81004 (719) 543-4016 Fax: (719) 583-6077 Email:kramerpa@ co.pueblo.co.us Thomas E. Henry, M.D. Chief Medical Examiner/ Coroner Denver City and County 660 Bannock Street Denver, CO 80204-4507 (303) 436-7711 (303) 436-7411 Fax: (303) 436-7709	
Connecticut	State Medical Examiner	State Chief Medical Examiner	• Due to criminal abortion, whether apparently self-induced or not. • Violent, whether apparently homicidal, suicidal, or accidental, including but not limited to deaths due to thermal, chemical, electrical, or radiation injury. • Sudden or unexpected deaths not caused by readily recognizable disease. • Under suspicious circumstances. • Where the body is to be cremated, buried at sea, or otherwise disposed of so as to be thereafter unavailable for examination. • Related to occupational disease or accident. • Related to disease which might constitute a threat to public health.	H. Wayne Carver, II, M.D. Chief Medical Examiner Office of the State Medical Examiner 11 Shuttle Road Farmington, CT 06032-1939 (860) 679-3980 Fax: (860) 679-1257 Email: H.Wayne.Carver@ po.state.ct.us	

State/ Territory	Type of System	Title	Deaths Investigated	Contact	Comment
Delaware	State Medical Examiner	State Chief Medical Examiner	• By violence, suicide, or casualty. • While under anesthesia. • By abortion or suspected abortion. • By poison or suspicion of poison. • Suddenly when in apparent health. • When unattended by a physician. • In prison or penal institution or police custody. • Resulting from employment. • From undiagnosed cause which may be related to a disease constituting a threat to public health. • In any suspicious or unusual manner. • If the body is unclaimed or to be cremated.	Richard Callery, M.D. State Chief Medical Examiner Office of the Chief Medical Examiner Department of Health and Social Services 200 South Adam Street Wilmington, DE 19801 (302) 577-3420 Fax: (302) 577-3416 Email: rcallery@state.de.us	
District of Columbia	Medical Examiner	Chief Medical Examiner	• By violence, whether apparently homicidal, suicidal, or accidental, including deaths due to thermal, chemical, electrical, or radiation injury. • Due to criminal abortion, whether apparently self-induced or not. • Under suspicious circumstances. • Of persons whose bodies are to be cremated, dissected, buried at sea, or otherwise disposed of so as to be thereafter unavailable for examination. • Sudden deaths not caused by readily recognizable disease. • Related to disease resulting from employment or to accident while employed. • Related to disease which might constitute a threat to public health.	Johnathan L. Arden, M.D. Chief Medical Examiner Office of the Chief Medical Examiner Building 27 1910 Massachusetts Avenue, Southeast Washington, DC 20003 (202) 698-9000 Fax: (202) 698-9100	

State/ Territory	Type of System	Title	Deaths Investigated	Contact	Comment
Florida	District Medical Examiner	District Medical Examiner	• By criminal violence. • By accident. • By suicide. • Suddenly, when in apparent good health. • Unattended by a physician or other recognized practitioner. • In a prison or penal institution. • In police custody. • In any suspicious or unusual circumstances. • By criminal abortion. • By poison. • By disease constituting a threat to public health. • By disease, injury, or toxic agent exposure resulting from employment. • When a body is brought into the State without proper medical certification. • When the body is to be cremated, dissected, or buried at sea.	Stephen J. Nelson, M.D., M.A., F.C.A.P. Chairman Medical Examiners Commission Florida Department of Law Enforcement P.O. Box 1489 Tallahassee, FL 32302-1489 (850) 410-8600 Fax: (850) 410-8621 Email:JNelsonMD@aol.com Vickie Marsey Program Admin. Medical Examiners Commission Same address as above (850) 410-8660 Fax: (850) 410-8621 Email: VickieMarsey@fdle.state.fl.us	
Georgia	Mixed: State Medical Examiner and County Coroners/Medical Examiners	County Coroner, State Medical Examiner, Regional Medical Examiner, County Medical Examiner, Local Medical Examiner	• By violence, suicide, or casualty. • Suddenly, when in apparent good health. • When unattended by a physician. • When an inmate of a State hospital or State, county, or city penal institution. • When ordered by a court having criminal jurisdiction. • After birth but before 7 years of age if the death is unexpected or unexplained.	Kris Sperry, M.D. State Medical Examiner Division of Forensic Sciences Georgia Bureau of Investigation P.O. Box 370808 Decatur, GA 30037-0808 (404) 244-2709 Fax: (404) 212-3047 Email: kris.sperry@gbi.state.ga.us	

State/ Territory	Type of System	Title	Deaths Investigated	Contact	Comment
Georgia (cont'd)			• In any suspicious or unusual manner, with particular attention to those persons 16 years of age and under. • As a result of an execution carried out pursuant to imposition of the death penalty.	Randy Hanzlick, M.D. Chief Medical Examiner, Fulton County Fulton County Medical Examiner Center 430 Pryor Street SW Atlanta, GA 30312 (404) 730-4400 Fax: (404) 730-4407 Email: RandyHanzlick@ mail.co.fulton.ga.us	
Guam	Territorial Medical Examiner	Chief Medical Examiner	• Violent, unusual, or unnatural deaths. This covers any death attributed to accident, suicide, homicide, criminal abortion, physical, mechanical, electrical, chemical, radiational, thermal, or related means. • All deaths under suspicious circumstances. • Sudden deaths in apparent health without obvious cause. • Deaths without medical attendance: 1. found dead without obvious or probable cause. 2. unattended at anytime by a licensed physician. 3. unattended by a physician during a terminal illness, particularly if such death appears unrelated to a disease previously diagnosed and treated. 4. fetal death attended by midwife. 5. deaths in prison, lock up, penitentiary, or juvenile justice facility. 6. deaths during or following an acute or unexplained syncope or coma. 7. during an acute or unexplained rapidly fatal illness which may be contagious to the public.	Aurelio A. Espinola, M.D. Chief Medical Examiner P.O. Box 7147 Tamuning, GU 96931 (671) 646-9363h	

State/ Territory	Type of System	Title	Deaths Investigated	Contact	Comment
Guam (cont'd)			8. all hospital DOAs (dead on arrival) and those dying within 24 hours after admission. • Bodies to be cremated, buried at sea, or shipped off island.		
Hawaii	Mixed: County Medical Examiners/ Coroners	County Coroner, Coroner Physician (titled Medical Examiner in Honolulu)	• As the result of violence. • As the result of any accident. • By suicide. • Suddenly when in apparent health. • When unattended by a physician. • In prison. • In a suspicious or unusual manner. • Within 24 hours after admission to a hospital or institution.	Kanthi Von Guenthner, M.D. Department of the Medical Examiner 835 Iwilei Road Honolulu, HI 96817 (808) 527-6777 Fax: (808) 524-8797	
Idaho	County Coroner	County Coroner	• As a result of violence whether apparently homicidal, suicidal, or accidental. • Under suspicious or unknown circumstances. • When not attended by a physician during his/her last illness and the cause of death cannot be certified by a physician.	Erwin Sonnenberg Idaho Coroners Association 5550 Morrishill Road Boise, ID 83706 (208) 364-2676 Fax: (208) 364-2685 Email: CRSONNEL@ ADAWEB.NET ˜	
Illinois	Mixed: County Medical Examiners/ Coroners	County Coroner, County Medical Examiner	• Sudden or violent death, whether apparently suicidal, homicidal, or accidental, including but not limited to deaths apparently caused or contributed to by thermal, traumatic, chemical, electrical or radiation injury, or a complication of any of them, or by drowning, suffocation, or motor vehicle accident. • Maternal or fetal death due to abortion. • Due to sex crime or crime against nature.	Jeff Lair Secretary Illinois Coroners and Medical Examiners Association P.O. Box 1261 Jacksonville, IL 62651 (217) 245-7423 Fax: (217) 479-4637 Email: coroner@ direcway.com	

State/ Territory	Type of System	Title	Deaths Investigated	Contact	Comment
Illinois (cont'd)			• Where the circumstances are suspicious, obscure, mysterious, or otherwise unexplained; and where, in the written opinion of the attending physician, the cause is not determined.		
			• Where addiction to alcohol or to any drug may have been a contributory cause.		
			• Where the decedent was not attended by a licensed physician.		
			• Occurring in State institutions or of wards of the State.		
			• If a child under 2 years dies suddenly or unexpectedly and circumstances concerning the death are unexplained.		
			• While being pursued, apprehended, or taken into custody by law enforcement officers or while in custody of any law enforcement agency.		
Indiana	County Coroner	County Coroner	• Has died from violence. • Has died by casualty. • Has died when apparently in good health. • Has died in an apparently suspicious, unusual, or unnatural manner. • Has been found dead.	Lisa Barker Executive Director Indiana Coroners Association 1643 West 800 South Romney, IN 47981 (877) 692-7284 Fax: (765) 538-2880 Email: ISCABarker@aol. com James St. Myer President Indiana Coroners Association 4109 Janney Ave. Muncie, IN 47305 (765) 289-0865 Fax: (765) 284-4606 Email: JSTM1079@aol. com	

State/ Territory	Type of System	Title	Deaths Investigated	Contact	Comment
Iowa	State Medical Examiner	State Medical Examiner, County Medical Examiner	• Violent, including homicidal, suicidal, or accidental. • Caused by thermal, chemical, electrical, or radiation injury. • Caused by criminal abortion including those self-induced, or by rape, carnal knowledge, or crimes against nature. • Related to disease thought to be virulent or contagious, which might constitute a public hazard. • Occurring unexpectedly or from unexplained causes. • Of a person confined in jail, prison, or correctional institution. • Where a physician was not in attendance at any time at least 36 hours preceding death, with the exception of pre-diagnosed terminal or bedfast cases for which the time period shall be extended to 20 days. • Where the body is not claimed by relatives or friends. • Where the identity of the deceased is unknown. • Of a child under the age of 2 years when sudden infant death syndrome is suspected or cause of death is unknown.	Julia C. Goodin, M.D., State Medical Examiner Office of the Iowa State Medical Examiner, Iowa Department of Public Health Lucas State Office Building, 5th Floor, 321 East 12th Street Des Moines, IA 50309 (515) 281-6726 Email: *jgoodin@raccoon. com*	
Kansas	District Coroner	District Coroner	• Death is suspected to have been the result of violence caused by unlawful means, or by suicide. • By casualty. • Suddenly when the decedent was in apparent health. • When the decedent was not regularly attended by a licensed physician.	Alan Hancock, M.D. President Kansas Coroners Association 9201 Parallel Parkway Kansas City, KS 66112-1598 (913) 299-1474 Fax: (913) 299-4931	
Kansas (cont'd)			• In any suspicious or unusual manner, or when in police custody, or when in a jail or correctional institution. • When the determination of the cause of death is held to be in the public interest.		

State/ Territory	Type of System	Title	Deaths Investigated	Contact	Comment
Kentucky	Mixed: State Medical Examiner and County Coroners/Medical Examiners	County Coroner, State Medical Examiner, District Medical Examiner	• Caused by homicide, violence, suicide, an accident, drugs, poison, motor vehicle, train, fire, explosion, drowning, illegal abortion, or unusual circumstances. • By criminal means. • Sudden infant death syndrome. • Child abuse. • Death in a person less than 40 years of age with no past medical history to explain the death. • Death occurring at a worksite. • Death in any mental institution. • Death in any prison, jail, or penal institution, or while decedent was in police custody. • When the death was sudden and unexplained or the decedent was unattended by a physician by more than 36 hours prior to death. • When skeletonized or extensively decomposed human remains are found. • When the body is to be cremated. • When circumstances of death result in a request by any responsible citizen for an investigation.	Tracey S. Corey, MD State Chief Medical Examiner Office of the State Medical Examiner Urban Government Center 810 Barrett Ave. Louisville, KY 40204 (502) 588-5587 Fax: (502) 852-1767 Email: TraceySCorey@ aol.com Dan Able Executive Director Office of the State Medical Examiner 100 Sower Blvd, Suite 202. Frankfort, KY 40601 (502) 564-4545 Fax: (502) 564-1699 Email: dan.able@ky.gov	

State/ Territory	Type of System	Title	Deaths Investigated	Contact	Comment
Louisiana	County Coroner	Parish Coroner	• When suspicious, unexpected, unusual or sudden. • By violence. • Due to unknown or obscure causes. • Where the body is found dead. • Without attending physician within 36 hours prior to the hour of death. • When abortion, whether self-induced or otherwise, is suspected. • Due to suspected suicide or homicide. • When poison is suspected. • From natural causes occurring in a hospital under 24 hours admission unless seen by a physician in the past 36 hours. • Following an injury or accident, either old or recent. • Due to drowning, hanging, burns, electrocution, gunshot wounds, stabs or cutting, lightning, starvation, radiation, exposure, alcoholism, addiction, tetanus, strangulation, suffocation, or smothering. • Due to trauma from whatever cause. • Stillborn deaths. • Due to criminal means. • If victim of alleged rape, carnal knowledge, or crime against nature. • By casualty. • In prison or while serving a sentence. • Due to virulent contagious disease that might be caused by or cause a public hazard (AIDS cases included).	Allen Herbert, M.D. President Louisiana Coroners' Association P.O. Box 747 Alexandria, LA 71309 (318) 473-6831 Fax: (318) 473-6832 Email: LSCA@speed-gate.net	

State/ Territory	Type of System	Title	Deaths Investigated	Contact	Comment
Maine	State Medical Examiner	State Chief Medical Examiner, Medical Examiner	• Violence of any kind. • Any cause where the death occurs suddenly while the person is in apparent good health. • Any cause where there is no attending physician capable of certifying the death as due to natural causes. • Poisoning, either chronic or acute. • Diagnostic or therapeutic procedures under circumstances indicating gross negligence or unforeseen clearly traumatic causes. • Any cause while the person is in custody or confinement, unless clearly certifiable by an attending physician as due to natural causes. • Disease or pathological process constituting a threat to public health, if the authority of the medical examiner is required to study the death adequately to protect the public health. • Deaths which may have been improperly certified or inadequately examined, including, but not limited to, bodies brought into the State under these circumstances. • In the case of a child under the age of 3 years, from any cause, including sudden infant death syndrome, unless the death is clearly due to a specific natural cause.	Margaret S. Greenwald, MD State Chief Medical Examiner State House, Station #37 Augusta, ME 04333 (207) 624-7180 Fax: (207) 624-7178 Email: margaret.greenwald@state.me.us	

State/ Territory	Type of System	Title	Deaths Investigated	Contact	Comment
Maryland	State Medical Examiner	State Chief Medical Examiner	• By violence, suicide, or casualty. • Suddenly when in apparent good health, or when unattended by a physician. • In any suspicious or unusual manner. • Fetuses, regardless of duration of pregnancy, if the mother is not attended by a physician at or after the delivery.	David Fowler, M.D. Acting State Chief Medical Examiner State of Maryland 111 Penn Street Baltimore, MD 21201 (410) 333-3226 Fax: (410) 333-3063 Email: OCMEMD@aol. com	
Massachusetts	State Medical Examiner	State Chief Medical Examiner, District Medical Examiner	• Where criminal violence appears to have taken place, regardless of the time interval between the incident and death, and regardless of whether such violence appears to have been the immediate cause of death, or a contributory factor. • By accident or unintentional injury, regardless of the time interval between the incident and death, and regardless of whether such violence appears to have been the immediate cause of death, or a contributory factor. • Suicide, regardless of the time interval between the incident and death. • Death under suspicious or unusual circumstances. • Death following an illegal abortion. • Death related to occupational illness or injury. • Death in custody, in any jail or correctional facility, or in any mental health or mental retardation facility. • Death where suspicion of abuse of a child, family or household member, elder person, or disabled person exists. • Death due to poison or acute or chronic use of drugs or alcohol.	Richard Evans, M.D. State Chief Medical Examiner Commonwealth of Massachusetts 720 Albany Street Boston, MA 02118 (617) 267-6767 Toll free in Massachusetts (800) 962-7877 Fax: (617) 266-6763 Email: richard.evans@ cme.state.ma.us	

State/ Territory	Type of System	Title	Deaths Investigated	Contact	Comment
Massachusetts (continued)			• Skeletal remains.		
			• Death associated with diagnostic or therapeutic procedures.		
			• Sudden death when the decedent was in apparent good health.		
			• Death within 24 hours of admission to a hospital or nursing home.		
			• Death in any public or private conveyance.		
			• Fetal death as defined by Section 202 of Chapter 111, where the period of gestation has been 20 weeks or more or where fetal weight is 350 grams or more.		
			• Pediatric deaths under and including the age of 18 years from any cause.		
			• Any person found dead.		
			• Death in an emergency treatment facility, medical walk-in center, day care center, or under foster care.		
			• Death occurring under such circumstances as the chief medical examiner shall prescribe in regulations promulgated pursuant to the provisions of Chapter 30 A.		
Michigan	County Medical Examiner	County Medical Examiner	• By violence. • When unexpected. • Without medical attendance during the 48 hours prior to the hour of death unless the attending physician, if any, is able to determine accurately the cause of death. • As the result of an abortion, whether self-induced or otherwise. • Of any prisoner in any county or city jail.	Dan Remick, M.D. President Michigan Association for Medical Examiners M2210 Medical Science I, 1301 Catherine Road Ann Arbor, MI 48109-0602 (734) 763-6454 Fax: (734) 763-6476 Email: remickd@umich. edu	

State/ Territory	Type of System	Title	Deaths Investigated	Contact	Comment
Minnesota	Mixed: County Medical Examiners/ Coroners	County Coroner, Medical Examiner (Hennepin, Ramsey Counties)	• Violent, whether apparently homicidal, suicidal, or accidental, including but not limited to deaths due to thermal, chemical, electrical, or radiation injuries. • Due to criminal abortion, whether apparently self-induced or not. • Under unusual or mysterious circumstances. • Of persons whose bodies are to be cremated, dissected, buried at sea, or otherwise disposed of so as to be thereafter unavailable for examination. • Of inmates of public institutions who are not hospitalized therein for organic diseases.	Andrew M. Baker, M.D. Medical Examiner Hennepin County 530 Chicago Avenue Minneapolis, MN 55415 (612) 215-6300 Fax: (612) 215-6330 Email: andrew.baker@ co.hennepin.mn.us Michael Rossman Executive Secretary Minnesota Coroners' and Medical Examiners' Association 530 Chicago Avenue Minneapolis, MN 55415 (612) 215-6300 FAX (612) 215-6330 Email: michael.rossman@co.hennepin.mn.us	
Mississippi	Mixed: State Medical Examiner and County Coroners/Medical Examiners	State Medical Examiner, County Medical Examiner (Coroner), County Medical Examiner Investigator (Coroner)	• Violent, including homicidal, suicidal, or accidental. • Caused by thermal, chemical, electrical, or radiation injury. • Caused by criminal abortion, including self-induced, or abortion related to or by sexual abuse. • Related to disease thought to be virulent or contagious which may constitute a public health hazard. • Unexpected or from an unexplained cause. • Of a person confined in a prison, jail, or correction institution (autopsy mandatory if prisoner was in custody of State Correctional System). • Of a person where physician was not in attendance within 36 hours preceding death, or in prediagnosed terminal or bedfast cases within 30 days. • Of a person where the body is not claimed by a relative or friend. • Of a person where the identity of the deceased is unknown.	Sam Howell Administrator Mississippi Crime Laboratory 1700 East Woodrow Wilson Avenue Jackson, MS 39216 (601) 987-1440 Fax: (601) 987-1445 Email: showell@mcl.state.ms.usl	

State/ Territory	Type of System	Title	Deaths Investigated	Contact	Comment
Mississippi (continued)			• Of a child under the age of 2 years where death results from an unknown cause or where the circumstances surrounding the death indicate that sudden infant death syndrome may be the cause of death (autopsy mandatory).		
			• Where a body is brought into this State for disposal and there is reason to believe either that the death was not investigated properly or that there is not an adequate certification of death.		
			• Where a person is admitted to a hospital emergency room unconscious and/or unresponsive, with cardiopulmonary resuscitative measures being performed, and dies within 24 hours of admission without regaining consciousness or responsiveness, unless a physician was in attendance within 36 hours preceding presentation to the hospital, or in cases in which the decedent had a prediagnosed terminal or bedfast condition, unless a physician was in attendance within 30 days preceding presentation to the hospital.		

State/ Territory	Type of System	Title	Deaths Investigated	Contact	Comment
Missouri	Mixed: County Medical Examiners/ Coroners	County Coroner, County Medical Examiner	Coroner—In any city of 700,000 or more inhabitants or in any county of the first or second class in which a Coroner is required, the Coroner must investigate all deaths where there is reason to believe that death was caused by criminal violence or following abortion. • By violence by homicide, suicide, or accident. • By criminal abortion, including those self- induced. • By some unforeseen occurrence and the deceased had not been attended by a physician during the 36-hour period preceding the death. • Occurring in any unusual or suspicious manner. Medical Examiner— • By violence by homicide, suicide, or accident. • By criminal abortion, including those self-induced. • By disease thought to be of a hazardous and contagious nature or which might constitute a threat to public health. • Suddenly when in apparent good health. • When unattended by a physician, chiropractor, or accredited Christian Science practitioner during the 36-hour period preceding the death. • While in the custody of the law. • While an inmate of a public institution. • Occurring in any unusual or suspicious manner.	Michael A. Graham, M.D. Medical Examiner City of St. Louis 1300 Clark Avenue St. Louis, MO 63103-2718 (314) 622-4971 Fax: (314) 622-4933 Email: GRAHAMMA@ slu.edu	

State/ Territory	Type of System	Title	Deaths Investigated	Contact	Comment
Montana	Mixed: State Medical Examiner and County Coroners/Medical Examiners	County Coroner, Chief State Medical Examiner, Associate Medical Examiner	Coroner— • Death caused or suspected to have been caused by an injury, either recent or remote in origin. • Death caused or suspected to have been caused by the deceased or any other person that was the result of an act or omission, including, but not limited to, a criminal or suspected criminal act; a medically suspicious death, unusual death, or death of unknown circumstances, including any fetal death; or an accidental death. • Death caused or suspected to have been caused by an agent, disease, or medical condition that poses a threat to public health. • Death occurring while the deceased was incarcerated in a prison or jail or confined to a correctional or detention facility owned and operated by the State or a political subdivision of the State. • Death occurring while the deceased was in the custody of, or was being taken into the custody of, a law enforcement agency or a peace officer. • Death occurring during or as a result of the deceased's employment. • Death occurring less than 24 hours after the deceased was admitted to a medical facility or if the deceased was dead upon arrival at a medical facility. • Death occurring in a manner that was unattended or unwitnessed and the deceased was not attended by a physician at any time in the 30-day period prior to death.	Gary Dale, M.D. State Medical Examiner State Crime Lab Division of Forensic Sciences 2679 Palmer Street Missoula, MT 59808 (406) 728-4970 Fax: (406) 549-1067 Email: gdale@state.mt.us Terry Bullis Secretary-Treasurer Montana Coroners' Association P.O. Box 318 Hardin, MT 59034-0318 (406) 665-1207 M.E. "Mickey" Nelson Lewis and Clark County Coroner 228 Broadway St. Helena, MT 59601-4263 (406) 442-7398 Fax: (406) 447-8298 Email: MNELSON@ co.lewis-clark.mt.us	

State/ Territory	Type of System	Title	Deaths Investigated	Contact	Comment
Montana (cont'd)			• If the dead body is to be cremated or shipped into the State and lacks proper medical certification or burial or transmit permits. • Death that occurred under suspicious circumstances. • Death that is the result of a judicial order. • Death that has occurred and no physician or surgeon licensed in Montana will sign a death certificate. ChiefState Medical Examiner—provides assistance and consultation to Coroners.		
Nebraska	County Coroner	County Coroner, County Coroner's Physician	• By criminal means or violence. • Homicide or suicide. • By drowning. • If sudden or unusual. • If drug-related. • If sudden infant death syndrome is suspected. • When involving the sudden and unexplained death of a child between the ages of 1 week and 3 years, and when neglect, violence, or any unlawful means are possible. • When death is not certified by attending physician. • When an individual has died while being apprehended by or while in the custody of a law enforcement officer or detention personnel. • Any suspicious, unexplained, or unattended death.	DeMaris Johnson Executive Director Nebraska County Attorneys Association Suite 203 1233 Lincoln Mall Lincoln, NE 68508 (402) 476-6047 Fax: (402) 476-2469	In Lancaster and Douglas Counties, physicians have been appointed to assist in the investigations and the signing of the medical portion of death certificates. In Douglas County, the County Attorney has appointed a medical examiner to investigate deaths by unnatural causes.

State/ Territory	Type of System	Title	Deaths Investigated	Contact	Comment
Nebraska (cont'd)					Coroner's physicians sign the death certificates. In Lancaster County, the County Attorney has retained a pathologist who assists in the investigation by performing autopsies and signing the medical portion of death certificates.
Nevada	District Coroner	District Coroner	• Unattended deaths. • Deaths wherein the deceased has not been attended by a physician in the 10 days before death. The coroner shall issue the certificate of death following consultation with a physician licensed to practice in the State. • Deaths related to or following known or suspected self-induced or criminal abortion. • Known or suspected homicide, suicide, or accidental death. • Deaths known or suspected as resulting in whole or in part from or related to accident or injury. • Deaths from drowning, fire, hanging, gunshot, stabbing, cutting, exposure, starvation, alcoholism, drug addiction, strangulation, or aspiration. • Deaths in whole or in part occasioned by criminal means. • Deaths in prison.	Lary Simms, M.D. Chief Medical Examiner Clark County 1704 Pinto Lane Las Vegas, NV 89106 (702) 455-3210 Fax: (702) 455-0416 Email: LSI@co.clark. nv.us Vernon O. McCarty Washoe County Coroner P.O. Box 11130 Reno, NV 89520 (702) 785-6114 Fax:(775) 785-1468 Email: VMCCARTY@ mail.co.washoe.nv.us	Although they have coroner systems by ordinance, Washoe County (Reno) and Clark County (Las Vegas) employ board-certified forensic pathologists who provide medical examiner services for their respective counties, as well as for other counties on a fee-for-service basis.

State/ Territory	Type of System	Title	Deaths Investigated	Contact	Comment
Nevada (cont'd)			• Deaths under such circumstances as to afford reasonable ground to suspect that the death was caused by the criminal act of another, or any deaths reported by physicians or other persons having knowledge of death for inquiry by the coroner.		
New Hampshire	State Medical Examiner	State Chief Medical Examiner, County Medical Examiner	• By violence or unlawful act. • In any suspicious, unusual, or unnatural manner. • In prison. • When unattended by a physician. • Suddenly when in apparent health, including those sudden and unexpected deaths of children under 3 years of age or when sudden infant death syndrome is suspected.	Thomas Andrew MD Chief Medical Examiner Office of State Chief Medical Examiner Suite 218 246 Pleasant Street Concord, NH 03301 (603) 271-1235 Fax: (603) 271-6308	
New Jersey	State Medical Examiner	State Medical Examiner, County Medical Examiner	• By violence whether apparently homicidal, suicidal, or accidental, including, but not limited to, deaths due to thermal, chemical, electrical, or radiation injury. • Deaths due to criminal abortion, whether apparently self-induced or not. • Not caused by readily recognizable disease, disability, or infirmity. • Under suspicious or unusual circumstances. • Within 24 hours after admission to a hospital or institution. • Of inmates in prison. • Of inmates of institutions maintained in whole or in part at the expense of the State or county where the inmate was not hospitalized therein for organic disease.	Faruk B. Presswalla, M.D. State Medical Examiner Office of the State Medical Examiner P.O. BOX 094 Trenton, NJ 08625-0094 (609) 896-8900 Fax: (609) 896-8697 Email: presswallaf@dcj. lps.state.nj.us	

State/ Territory	Type of System	Title	Deaths Investigated	Contact	Comment
New Jersey (cont'd)			• From causes which might constitute a threat to public health. • Related to disease resulting from employment or to accident while employed. • Sudden or unexpected deaths of infants and children under 3 years of age. • Fetal deaths occurring without medical attendance.		
New Mexico	State Medical Examiner	State Medical Investigator, District Medical Investigator	• Sudden, violent, or untimely. • Found dead and the cause of death is unknown or obscure. • If caused by criminal act or omission.	Ross E. Zumwalt, M.D. State Chief Medical Investigator Office of the Medical Investigator State of New Mexico MSC11 6030 1 University of New Mexico Albuquerque, NM 87131-0001 (505) 272-3053 Fax: (505) 272-0727 Email:RZumwalt@salud. unm.edu	
New York	Mixed: County Medical Examiners/ Coroners	County Coroner, County Coroner's Physician, County Medical Examiner (in counties abolishing coroner system)	• By violence, whether criminal violence, suicide, or casualty. • Caused by unlawful act or criminal neglect. • Occurring in a suspicious, unusual, or unexplained manner. • Caused by suspected criminal abortion. • While unattended by a physician, so far as can be discovered, or where no physician is able to certify the cause of death as provided in public health law and in form as prescribed by the Commissioner of Health can be found. • Of a person confined in a public institution other than a hospital, infirmary, or nursing home. • Death occurring to an inmate of a correctional facility.	Charles S. Hirsch, M.D. Chief Medical Examiner City of New York (New York, Bronx, Kings, Queens, and Richmond Counties) 520 First Avenue, Room 134 New York, NY 10016 (212) 447-2034 Fax: (212) 447-2744	In some counties, there are multiple coroners, each having equal authority. In Lewis, Madison, and Oswego Counties, State law requires that the District Attorney serve as Coroner.

State/ Territory	Type of System	Title	Deaths Investigated	Contact	Comment
North Carolina	Mixed: State Medical Examiner and County Coroners/Medical Examiners	Chief Medical Examiner, County Medical Examiner, (Acting) Medical Examiner	• Homicide. • Suicide. • Trauma-related. • Accidental. • Disaster related. • Violence related. • Unknown, unnatural, unusual, or suspicious circumstances. • In police custody, jail, prison, or correctional institution. • Poisoning or suspicion of poisoning. • Possible public health hazard (such as acute contagious disease or epidemic). • Deaths during surgical or anesthetic procedure. • Sudden unexpected deaths which are not reasonably related to known previous disease. • Deaths without medical attendance, as defined by statute.	John D. Butts, M.D. Chief Medical Examiner State Department of Health and Human Services Office of the Chief Medical Examiner Chapel Hill, NC 27599-7580 (919) 966-2253 Fax: (919) 962-6263 Email: jbutts@ocme.unc.edu	North Carolina has coroners in some counties who work closely with the State Medical Examiner.
North Dakota	County Coroner	County Coroner	• Generally, deaths occurring by unlawful means or without medical attendance. • In counties with more than 8,000 population. • As a result of criminal or violent means. • By casualty or accident. • By suicide. • Suddenly when in apparent good health. • In a suspicious or unusual manner. • Occurring without medical attendance. • When the Workers' Compensation Board deems it necessary under the Crime Victims Reparation Act.	Beverly R. Wittman Deputy State Registrar and Director Division of Vital Records North Dakota State Department of Health 600 East Boulevard Bismarck, ND 58505 (701) 328-4508 Fax:(701) 328-1850 Email: BWITTMAN@ state.nd.usNorth Dakota	The office of Coroner is abolished in counties adopting the County Manager form of government, with the County Manager or Sheriff assuming the duties of the Coroner. However, if these duties conflict with those performed by the Sheriff, the county State's Attorney assumes the duties of the coroner.

State/ Territory	Type of System	Title	Deaths Investigated	Contact	Comment
Northern Marina Islands	Not available	Not Available	• Not available.	Orana Castro Director Office of Vital Statistics Commonwealth Trial Courts P.O. Box 307 Saipan, MP 96950 (670) 234-6401	
Ohio	Mixed: County Medical Examiners/ Coroners	County Coroner	• As a result of criminal or other violent means. • By casualty. • By suicide. • Suddenly when in apparent health. • In any suspicious or unusual manner. • Threat to public health.	David P. Corey Executive Director Ohio State Coroners Association 6161 Busch Blvd Suite #87 Columbus, OH 43229-2508 614-262-OSCA Fax: 614-888-9767 Email: info@osca.net	Coroners are not allowed to actively practice law while in office. Summit County has a County Medical Examiner system.
Oklahoma	State Medical Examiner	State Chief Medical Examiner, County Medical Examiner	• By violence, whether apparently homicidal, suicidal, or accidental, including, but not limited to, death due to thermal, chemical, electrical, or radiation injury. • Due to criminal abortion, whether apparently self-induced or not. • Under suspicious, unusual, or unnatural circumstances. • Related to disease that might constitute a threat to public health. • Unattended by a licensed medical or osteopathic physician for a fatal or potentially fatal illness. • Of persons after unexplained coma. • That are medically unexpected and occur in the course of a therapeutic procedure. • Of any inmates occurring in any place of penal incarceration. • Of persons whose bodies are to be cremated, buried at sea, transported out of State, or otherwise made ultimately unavailable for pathological study.	Fred B. Jordan, M.D. Chief Medical Examiner 901 North Stonewall Oklahoma City, OK 73117 (405) 239-7141 Fax: (405) 239-2430 Email: Medical_ Examiner@ocmeokc. state.ok.us	

State/ Territory	Type of System	Title	Deaths Investigated	Contact	Comment
Oregon	State Medical Examiner	State Medical Examiner, County Medical Examiner	• Apparently homicidal, suicidal, or occurring under suspicious or unknown circumstances. • Resulting from the unlawful use of dangerous or narcotic drugs or the use or abuse of chemicals or toxic agents. • Occurring while incarcerated in any jail, correction facility, or in police custody. • Apparently accidental or following an injury. • By disease, injury, or toxic agent exposure during or arising from employment. • While not under the care of a physician during the period immediately previous to death. • Related to disease which might constitute a threat to the public health.	Karen Gunson, M.D. State Medical Examiner Medical Examiner Division Oregon Department of State Police 301 Northeast Knott Street Portland, OR 97212-3092 (503) 988-3746 Fax: (503) 280-6041 Email: karen.gunson@ state.or.us Eugene Gray Forensic Administrator Medical Examiner Division Oregon Department of State Police 301 Northeast Knott Street Portland, OR 97212-3092 (503) 280-6061 Fax: (503) 280-6041 Email: eugene.gray@ state.or.us	
Pennsylvania	Mixed: County Medical Examiners/ Coroners	County Coroner, County Medical Examiner	• Sudden deaths not caused by readily recognizable disease, or wherein the cause of death cannot be properly certified by a physician on the basis of prior (recent) medical attendance. • Deaths occurring under suspicious circumstances, including those where alcohol, drugs, or other toxic substances may have had a direct bearing on the outcome. • Deaths occurring as a result of violence or trauma, whether apparently homicidal, suicidal, or accidental (including, but not limited to, those due to mechanical, thermal, chemical, electrical or radiational injury, drowning, cave-ins, and subsidences).	Dennis Kwiatkowski Secretary Treasurer Pennsylvania State Coroners Association 110 Franklin Street Suite 500 Johnstown, PA 15901 (814) 535-622 Fax: (814) 539-9057 Email: coroner@ co.cambria.pa.us	

State/ Territory	Type of System	Title	Deaths Investigated	Contact	Comment
Pennsylvania (cont'd)			• Any death in which trauma, chemical injury, drug overdose, or reaction to drugs or medication or medical treatment was a primary or secondary, direct or indirect, contributory, aggravating, or precipitating cause of death. • Operative and perioperative deaths in which the death is not readily explainable on the basis of prior disease • Any death wherein the body is unidentified or unclaimed. • Deaths known or suspected as due to contagious disease and constituting a public hazard. • Deaths occurring in prison or a penal institution or while in the custody of the police. • Deaths of persons whose bodies are to be cremated, buried at sea, or otherwise disposed of so as to be thereafter unavailable for examination. • Sudden infant death syndrome. • Stillbirths.		
Puerto Rico	Territorial Medical Examiner	Director	• As a result of criminal acts or acts that are suspected as such. • As a result of any accident or act of violence or any subsequent act, regardless of its nature or time interval between said acts and death, if there is a reasonable doubt that there might have been a relation between said accident or act of violence and death. • As a result of poisoning or suspicion of poisoning. • Occurring in custody of the police or officers of the law, while in prison, or as a result of sickness or injury occurring while in prison, or suspicion thereof.	Lyvia A. Alvarez, M.D. Director Institute of Forensic Sciences Call Box 11878 Caparra Height Station San Juan, PR 00922 (809) 765-0615 (809) 765-4880	

State/Territory	Type of System	Title	Deaths Investigated	Contact	Comment
Puerto Rico (cont'd)			• As a result of or in relation to the occupation of the deceased.		
			• Due to acute intoxication with alcohol, narcotics, or any type of drug or controlled substance or suspicion of such.		
			• Due to suicide or suspected suicide.		
			• When in process of an autopsy that was not originally considered medicolegal, the pathologist discovers any clue, or any suspicion arises to indicate that such death could have occurred due to the commission of any crime.		
			• Occurring suddenly or unexpectedly, while the person was enjoying relative or apparent good health.		
			• Occurring during or after an abortion or childbearing.		
			• When the physician who attended said person while living cannot reasonably establish that the death was due to natural causes.		
			• Occurring during or after surgical, diagnostic, or therapeutic procedures or where the deceased was under anesthesia or recovering from it.		
			• Occurring during the course of an illness, if there is a suspicion that factors extraneous to said illness could have contributed to the death.		
			• Occurring in a convalescent home, asylum, or similar institution, whether it be commonwealth, municipal, or private.		
			• Occurring to a person who has had a contagious disease that could constitute a threat to public health.		

State/ Territory	Type of System	Title	Deaths Investigated	Contact	Comment
Puerto Rico (cont'd)			• Occurring within 24 hours after the admission of the patient to a hospital, clinic, or asylum, whether it be commonwealth, municipal, or private, whenever the death cannot be attributed to natural causes.		
			• Occurring during hospitalization in a psychiatric institution, whether it be commonwealth, municipal, or private, except in cases of death due to childbirth, duly certified by a physician.		
			• Death caused by physical force such as electricity, heat, cold, radiation, or the effect of chemical products.		
			• Any death due to malnutrition, abandonment, or exposure to the elements, or as a result of negligence.		
			• When the corpse is to be cremated, dissected, or it is to be disposed of in such a way that it will not be available subsequently for examination, regardless of how the death occurred.		
			• When the prosecutor or trial judge investigating the death requests an autopsy.		
Rhode Island	State Medical Examiner	State Medical Examiner	• By homicide, suicide or casualty. • Due to a criminal abortion. • Due to an accident involving lack of due care on the part of a person other than the deceased. • Which is the immediate or remote consequences of any physical or toxic injury incurred while the deceased person was employed.	Elizabeth A. Laposata, M.D. State Medical Examiner Office of State Medical Examiner Rhode Island Department of Health 48 Orms Street Providence, RI 02904 (401) 222-5500 Fax: (401) 222-5517	

State/ Territory	Type of System	Title	Deaths Investigated	Contact	Comment
Rhode Island (cont'd)			• Due to the use of addictive or unidentifiable chemical agents. • Due to an infectious agent capable of spreading an epidemic within the State. • When unattended by a physician.		
South Carolina	Mixed: County Coroner and Medical Examiner	County Coroner, County Medical Examiner	• By violence. • By suicide. • When in apparent good health. • When unattended by a physician. • In any unusual or suspicious manner. • While an inmate in a penal or correctional institution. • As a stillbirth, medically unattended.	Debbie Johnson Executive Director South Carolina Coroners Association 301 University Ridge, Suite 2300 Greenville, SC 29601 (864) 467-7446 Fax: (864) 467-7469	In counties with populations of 100,000 or more, the governing body of that county can chose to have a Medical Examiner system.
South Dakota	County Coroner	County Coroner	• By unnatural means, including all deaths of accidental, homi-cidal, suicidal, and undetermined manner, regardless of suspected criminal involvement in the death. • Identity of victim is unknown or the body is unclaimed. • Inmates of any state, county or municipally operated correctional facility, mental institution or special school. • Those believed to represent a public health hazard. • Children under 2 years of age resulting from unknown cause or if circumstances suggest sudden infant death syndrome as the cause. • Natural deaths if the decedent is not under the care of a physician or the decedent's physician does not feel qualified to sign the death certificate.	Kathlene A. Mueller Manager, Office of Data, Statistics and Vital Records State Department of Health 600 East Capitol Avenue Pierre, SD 57501-3182 (605) 773-3361 Fax: (605) 773-5683 Email: KATHIMU@doh. state.sd.us	

State/ Territory	Type of System	Title	Deaths Investigated	Contact	Comment
Tennessee	State Medical Examiner	State Chief Medical Examiner, Deputy Chief Medical Examiner, Assistant Chief Medical Examiner, County Medical Examiner	• From sudden violence. • By casualty. • By suicide. • Suddenly when in apparent health. • When found dead. • In prison. • In any suspicious, unusual, or unnatural manner. • Where the body is to be cremated. • For workers compensation claims if cause of death is obscure or disputed.	Bruce Levy, M.D. State Chief Medical Examiner State of Tennessee Center for Forensic Medicine 850 R.S. Gass Blvd. Nashville, TN 37216-2640 (615) 743-1800 Fax: (615) 743-1890 Email: blevy@forensic-med.com	
Texas	Mixed: County Medical Examiners/ Coroners	County Justice of the Peace, County/ District Medical Examiner	Justice of the Peace— • In prison or jail. • When a person is killed, or from any cause dies an unnatural death, except under sentence of the law. • In the absence of one or more good witnesses. • When found dead, the circumstances of death unknown. • When the circumstances are such as to lead to suspicion of unlawful means. • By suicide or suspected suicide. • When unattended by a duly licensed and practicing physician and the local health officer or registrar required to report the cause of death does not know the cause of death. • When the attending physician(s) cannot certify the cause of death. Medical Examiner— • Same as above, and in addition, any death within 24 hours after admission to a hospital or institution.	Vincent J.M. DiMaio, M.D. State Chief Medical Examiner Bexar County Forensic Science Center 7337 Louis Pasteur San Antonio, TX 78229-4565 (210) 615-2100	

State/ Territory	Type of System	Title	Deaths Investigated	Contact	Comment
U.S. Virgin Islands	Territorial Medical Examiner	Medical Examiner	• Unnatural deaths as prescribed by law. • From violence, whether criminal, suicide or casualty. • Unlawful act or criminal neglect. • Suspicious, unusual, or unexplained manner. • Suspect criminal abortion. • Death unattended by physician or where cause of death unable to be certified as provided by law. • When confined to a public institution other than hospital, infirmary, or nursing home. • As prescribed by the Governor or Attorney General.	James Glenn, M.D. Medical Examiner Office of the Medical Examiner U.S. Virgin Islands Department of Justice Toro Building 3008 Golden Grove Christiansted St. Croix, VI 00820 (809) 778-6311 Francisco Landron, M.D. Medical Examiner Office of the Medical Examiner U.S. Virgin Islands Department of Justice 8050 Kronprindsens Gade St. Thomas, VI 00802 (809) 744-5666 ext. 663 (809) 776-8311	
Utah	State Medical Examiner		• By violence, gunshot, suicide, or accident (except highway accidents). • Suddenly when in apparent health. • When unattended (not seen by a physician within 30 days). • Under suspicious or unusual circumstances. • Resulting from poisoning or overdose of drugs. • Resulting from diseases that may constitute a threat to public health. • Resulting from disease, injury, toxic effect, or unusual exertion incurred within the scope of the deceased's employment. • Due to sudden infant death syndrome.	Todd C. Grey, M.D. State Medical Examiner Utah Department of Health 48 North Medical Drive Salt Lake City, UT 84113 (801) 584-8410 Fax: (801) 584-8435 Email:hldels.tgrey@state.ut.usa	

State/Territory	Type of System	Title	Deaths Investigated	Contact	Comment
			• Resulting when the deceased was in prison, jail, police custody, the State hospital, or a detention or medical facility operated for the treatment of mentally ill or emotionally disturbed or delinquent persons.		
			• Associated with diagnostic and therapeutic procedures.		
			• Involving questions of civil liability, in accordance with provisions of the Worker's Compensation Act.		
Vermont	State Medical Examiner	Chief Medical Examiner, Regional Medical Examiner	• From violence. • Suddenly when in apparent good health. • When unattended by a physician. • By casualty. • By suicide. • As a result of injury. • When in jail, prison, or any mental institution. • In any unusual, unnatural, or suspicious manner. • In circumstances involving a hazard to public health, welfare, or safety.	Paul L. Morrow, M.D. State Chief Medical Examiner Department of Health Baird 1 111 Colchester Ave. Burlington, VT 05401 (802) 863-7320 Fax: (802) 863-7265 Email: pmorrow@vdh.state.vt.us	

State/ Territory	Type of System	Title	Deaths Investigated	Contact	Comment
Virginia	State Medical Examiner	State Chief Medical Examiner, County/ City Medical Examiner	• By trauma, injury, violence, poisoning, accident, suicide, or homicide. • Suddenly when in apparent good health. • When unattended by a physician. • In jail, prison, other correctional institution or in police custody. • Suddenly as an apparent result of fire. • In any suspicious, unusual, or unnatural manner. • The sudden death of any infant less than 18 months of age whose death is suspected as due to SIDS. • When the body shall be cremated or buried at sea. • Fetal death not attended by a physician.	Marcella Farinelli Fierro, M.D. State Chief Medical Examiner Department of Health Office of the State Chief Medical Examiner 400 East Jackson Street Richmond, VA 23219-3694 (804) 786-3174 Fax: (804) 371-8595 Email: mfierro@vdh. state.va.us	
Washington	Mixed: County Medical Examiners/ Coroners	County Coroner, County Medical Examiner	• Those in which the Coroner suspects that the death was unnatural, or violent, or resulted from unlawful means, or from suspicious circumstances, or was of such a nature as to indicate the possibility of death by the hand of the deceased or through the instrumentality of some other person. • Those occurring suddenly when in apparent good health and without medical attendance within 36 hours preceding death. • Those resulting from unknown or obscure causes. • Those occurring within 1 year following an accident. • Those as a result of any violence whatsoever. • Those resulting from a known or suspected abortion, whether self-induced or otherwise.	Richard Harruff, M.D., Ph.D. Chief Medical Examiner, King County 325 Ninth Avenue HMC Box 35792 Seattle, WA 98104 (206) 731-3232 Fax: (206) 731-8555 Email: richard.harruff@ metrokc.gov Dan Blasdel, President Washington Association of Coroners and Medical Examiners 1016 North 4th Avenue Pasco, WA 99301 (509) 546-5885 Fax: (509)546-5812, page (509) 530-6906 Email: dblasdel@3-cities. com	

State/ Territory	Type of System	Title	Deaths Investigated	Contact	Comment
Washington (cont'd)			• Those apparently resulting from drowning, hanging, burns, electrocution, gunshot wounds, stabs or cuts, lightning, starvation, radiation, exposure, alcoholism, narcotics or other addictions, tetanus, strangulation, suffocation, or smothering. • Those due to premature birth or stillbirth. • Those due to virulent or suspected. contagious disease which may be a public health hazard. • Those resulting from alleged rape, carnal knowledge, or sodomy. • Those occurring in a jail or prison. • Those in which a body is found dead or is not claimed by relatives or friends. • Industrial deaths when cause of death is unknown and investigation is requested by the Department of Labor and Industries.		
West Virginia	State Medical Examiner	State Chief Medical Examiner, Deputy Chief Medical Examiner, County Medical Examiner	• From violence or suspected violence, or where natural disease cannot be assumed. • When unattended by a physician. • When during incarceration, protective custody, as a ward of the State, or associated with police intervention. • From disease or environmental condition which might constitute a threat to the public health. • When in any suspicious, unusual, or unnatural manner. • In deaths thought to be due to, or associated with, suspected abuse or neglect.	James Kaplan, M.D. State Chief Medical Examiner State of West Virginia Office of the Chief Medical Examiner 701 Jefferson Road South Charleston, WV 25309 (304) 558-3920 Fax: (304) 558-7886 Email: jkaplan@wvdhhr. org	

State/ Territory	Type of System	Title	Deaths Investigated	Contact	Comment
Wisconsin	Mixed: County Medical Examiners/ Coroners	County Coroner, County Medical Examiner	• If circumstances are unexplained, unusual, or suspicious. • By homicide or manslaughter, including death resulting from reckless conduct, negligent control of a vicious animal, or negligent use of a firearm. • By suicide. • Following an abortion. • By poisoning, whether homicidal, suicidal, or accidental. • Following accidents, whether the injury is or is not the primary cause. • When a physician or accredited practitioner of a bona fide religious denomination relying upon prayer or spiritual means for healing was not in attendance within 30 days preceding death, or if the deceased was not being treated for the condition causing death. • When a physician refuses to sign the death certificate. • When a physician cannot be obtained to sign the medical certification of death. • At the request of the Worker's Compensation Department.	Tom Terry, M.D. President Wisconsin Coroner and Medical Examiner Association 730 Wisconsin St Racine, WI 53403 (262) 636-3303 Fax: (262) 636-3728 Email: TOMT@racineco. com	Counties with a Medical Examiner include those having a population of 500,000 or more, and those that have chosen to have a Medical Examiner system.

State/ Territory	Type of System	Title	Deaths Investigated	Contact	Comment
Wyoming	County Coroner	County Coroner	• Violent or criminal action. • Apparent suicide. • Accident. • Apparent drug or chemical overdose or toxicity. • The deceased was unattended or had not seen a physician within 6 months prior to death. • Apparent child abuse causes. • The deceased was a prisoner, trustee, inmate, or patient of any county or state corrections facility or state hospital. • If the cause is unknown.	Donald B. Pierson Executive Director Wyoming Peace Officers Standards and Training Commission 1710 Pacific Avenue Cheyenne, WY 82002 (307) 777-7718 Fax: (307) 638-9706 Email: wypost@sisna.com James W. Thorpen, M.D. Chairman Wyoming Board of Coroner Standards County Building 200 North Center St, Suite 10 Casper, WY 82601-1949 (307) 235-9458 Fax: (307) 235-9608 Email: coroner@natrona.net	

SOURCE: all information was obtained from http://www.cdc.gov/epo/dphsi/mecisp/death_investigation.htm

■ Appendix B: Niosh Firefighter Fatality Reports—Autopsy Information

The NIOSH Fire Fighter Fatality Investigation and Prevention Program (FFFIPP) investigates firefighter line-of-duty fatalities. The purpose of this investigative program is, according to the FFFIPP Web site (http://www.cdc.gov/niosh/fire/implweb.html), threefold:

> "Better define the magnitude and characteristics of line-of-duty deaths among fire fighters"
>
> "Develop recommendations for the prevention of deaths and injuries"
>
> "Disseminate prevention strategies to the fire service"

The FFFIPP reports include specific details about the circumstances surrounding the fatalities as well as medical information including cause of death and pertinent autopsy results, when available.

This summary concentrates on investigations completed in the years 2000 through 2006; a few of the year 2000 investigations related to incidents in 1998 and 1999. The following table includes for each investigation the NIOSH Report Number, Date of Incident, Title, and Cause of Death, and includes notations about whether an autopsy was performed and autopsy findings. A link to each report's PDF file, available online, is provided. When the death certificate and autopsy listed differing causes of death, both are included in the table.

Most reports specifically mention the autopsy (or lack thereof). However, about one-third of the reports did not mention the autopsy outright. For some of those cases, postmortem toxicological specimen analysis, typically for carboxyhemoglobin level, was noted; however, the specimen was not necessarily obtained as part of an autopsy.

Approximately 280 FFFIPP reports were reviewed for recommendations or comments related to autopsy procedures and protocols. The most common comment specific to autopsy protocols was the recommendation to "perform autopsies on all on-duty fire fighter fatalities." This was noted on about two dozen cases involving cardiovascular deaths. No other recommendations were made for specific protocols or procedures for the autopsy itself, or for the postautopsy storage of tissue samples, etc. However, carboxy-hemoglobin levels were mentioned numerous times with a comment that postmortem levels were artificially low in a case in which the victim had received oxygen therapy prior to death.

TABLE: SUMMARY OF FFFIPP REPORTS

NIOSH REP. #	DATE OF INCIDENT	Title	CAUSE OF DEATH	AUTOPSY	AUTOPSY COMMENTS	PDF LINK
F2007-01	Dec 30, 2006	Career firefighter dies and chief is injured when struck by 130-foot awning that collapses during a commercial building fire—Texas	Blunt-force trauma	Not mentioned	None	F2007-01PDF
F2006-26	Aug 13, 2006	Career engineer dies and firefighter injured after falling through floor while conducting a primary search at a residential structure fire—Wisconsin	Smoke inhalation and thermal burns	Not mentioned	None	F2006-26PDF
F2006-22	Apr 8, 2006	Volunteer firefighter dies after being struck by a shackle on a recoiling tow rope—South Dakota	Blunt-force head trauma	Not mentioned	None	F2006-22PDF
F2006-19	May 14, 2006	Career lieutenant dies in residential structure fire—Colorado	Complications from hypoxic encephalopathy (lack of oxygen to the brain) due to apparent smoke inhalation	Not mentioned	None	F2006-19PDF
F2006-10	Mar 1, 2006	Volunteer firefighter is killed and another volunteer firefighter is injured at a wildland/urban interface fire—Oklahoma	Complications from the burns he received to over 45% of his body	Yes	None	F2006-10PDF
F2006-06	Nov 22, 2005	Volunteer firefighter dies in tanker rollover crash—Texas	Open skull fracture resulting from an automobile accident	Not mentioned	None	F2006-06PDF
F2005-28	Aug 6, 2005	One career firefighter dies and two are injured in apparatus crash—California	Multiple blunt-force trauma to the torso	Yes	None	F2005-28PDF

NIOSH REP. #	DATE OF INCIDENT	Title	CAUSE OF DEATH	AUTOPSY	AUTOPSY COMMENTS	PDF LINK
F2006-25	Jul 26, 2006	Junior volunteer firefighter dies and three volunteer firefighters are injured in a tanker crash—Alabama	Blunt force trauma with head and chest injuries	Not mentioned	None	⤓PDF
F2006-21	Mar 09, 2006	Career lieutenant suffers sudden cardiac death at his home after finishing his shift—Tennessee	Hypertensive heart disease	Yes	• Cardiomegaly (enlarged heart): heart weight of 441 grams (g) (normal weight is <400 g) • LVH – left ventricle thickness was 1.5-2.0 centimeters (cm) (normal thickness is 0.6-1.1 cm)1 • Myocardial perivascular and patchy interstitial fibrosis • Mild bilateral dilatation of left and right ventricles (no measurements were listed) • Moderate to severe coronary artery disease (CAD) with 75% narrowing of the left anterior descending (LAD) artery • Mild aortic sclerosis • No evidence of thrombosis (blood clots) in his coronary arteries • No evidence of a pulmonary embolus	⤓PDF
F2006-20	Apr 16, 2006	Career airport fire apparatus operator suffers sudden cardiac death at his station after exercising—Georgia	Acute Respiratory Distress Syndrome due to an acute myocardial infarction (MI)	Yes	• Acute MIo Massive infarct of the left ventricle, extending from the apex to the base of the heart – Focal acute extension of the infarct – Stent in place, proximal left anterior descending coronary artery, with a luminal thrombus• Atherosclerotic CADo Moderate to severe atherosclerosis of distal left anterior descending artery – Moderate atherosclerosis of right coronary artery – Moderate atherosclerosis of circumflex artery • Cardiomegaly (enlarged heart): heart weight of 600 grams (g) (the normal weight given in the autopsy records is 250-350 g) • No pulmonary embolus	⤓PDF

NIOSH REP. #	DATE OF INCIDENT	Title	CAUSE OF DEATH	AUTOPSY	AUTOPSY COMMENTS	PDF LINK
F2006-18	Mar 31, 2006	Firefighter/emergency medical technician (ff/emt) suffers sudden death while on-duty—South Carolina	Per death certificate: "probable cardiac arrhythmia," due to "seizures," due to "epilepsy." Per autopsy findings: "sudden unexplained ventricular arrhythmia or possible seizure."	Yes	• Normal sized heart, 350 grams (normal < 400 grams) • No plaque, atherosclerosis, or blockages in any of the coronary arteries • Two coronary arteries (the left anterior descending and the right coronary artery) had "small diameters within their distal distribution" • No microscopic evidence of cardiomyopathy (a medical condition that is associated with an increased risk of sudden cardiac death) • No evidence of a pulmonary embolus (blood clot in the lung arteries) • No evidence of an intra-cranial hemorrhage (stroke) • Negative blood drug test for illegal • Positive blood drugs test for methobarbitol and phenobarbital. The phenobarbital level of 8.9 micrograms per milliliter (µg/mL) was sub-therapeutic for a seizure disorder (10-30 µg/mL). A blood level for dilantin was not conducted. These autopsy findings did not point to a definitive cause of death, therefore the county coroner concluded the FF/EMT most likely died of a "sudden unexplained ventricular arrhythmia or possible seizure."	PDF

F2006-17	May 03, 2006	Firefighter suffers heart attack during firefighting operation and dies 40 days later—Georgia	"Hemodynamic failure" due to "healing and remote myocardial infarctions (mis)" due to "atherosclerotic coronary artery disease (CAD)"	Yes	• Atherosclerotic CAD
					   – Normal-sized heart at 400 grams
					   – Old (healed) MI in the lateral portion of the left ventricle
					   – Recent MI in the anteroseptal portion of the left ventricle
					   – Atherosclerotic lesions (50%—60% blockage) in most of the coronary arteries
					   – Recent plaque hemorrhage and rupture of the proximal portion of the LAD
					 • No valve abnormalities
					 • No chamber dilation or hypertrophy
						PDF

NIOSH REP. #	DATE OF INCIDENT	Title	CAUSE OF DEATH	AUTOPSY	AUTOPSY COMMENTS	PDF LINK

F2006-16	Jul 07, 2005	Fire apparatus operator suffers sudden cardiac death after responding to 12 calls—Georgia	"Cardiac dysrhythmia due to atherosclerotic coronary artery disease (CAD)"	Yes	• Atherosclerotic CAD
					  – Variable calcific atherosclerosis of the left anterior descending coronary artery with a 90% blockage
					  – Variable calcific atherosclerosis of the right coronary artery with a 40% blockage
					  – Cardiomegaly (enlarged heart: heart weighed 660 grams
					  – Left ventricular hypertrophy
					• Microscopic examination of the left ventricle revealed "patchy transmural myocardial fibrosis." "Similar but lesser changes were noted in the right ventricular myocardium, where fibrosis was primarily perivascular."
					• No evidence of thromboemboli in the pulmonary arteries
					• Negative drug and alcohol tests
						⬇PDF

F2006-15	Apr 03, 2006	Assistant chief suffers a stroke during training and dies—Texas	Brain death due to brain aneurysm	No	Perform an autopsy on all duty-related firefighter fatalies	⬇PDF
F2006-13	Jan 14, 2006	Volunteer firefighter suffers sudden cardiac death about 50 minutes after fighting a grass fire—Kansas	Probable heritable cardiac arrhythmia (Brugada Syndrome)	Yes	• Probable heritable cardiac arrhythmia (Brugada Syndrome) • Cardiomegaly (heart weighed 478 grams [g]; normal range is 261-455 g) • Coronary arteries free of significant atherosclerosis or thrombus • No pulmonary embolus	⬇PDF

NIOSH REP. #	DATE OF INCIDENT	Title	CAUSE OF DEATH	AUTOPSY	AUTOPSY COMMENTS	PDF LINK
F2006-12	Nov 15, 2004	Career battalion chief suffers sudden cardiac death at his desk—Kansas	Atherosclerotic cardiovascular disease (CVD)	Yes	• Cardiomegaly (enlarged heart): heart weighed 534 grams • Left ventricular hypertrophy (LVH) • Essentially normal endocardium, myocardium, and epicardium tissue. • Two vessel coronary artery disease (CAD) with 50%-75% narrowing of the left anterior descending and right coronary arteries by atherosclerotic plaque • No evidence of thrombosis • Mild fibrous thickening of the cusps of the aortic valve • No evidence of a pulmonary embolus.	⟲PDF
F2006-09	Jan 28, 2006	Firefighter dies after performing overhaul at a fire in a three-story dwelling—Pennsylvania	"Arteriosclerotic cardiovascular disease (CVD)" as the immediate cause of death and "hypertensive cardiomyopathy" as a significant condition	Yes	• Cardiomegaly (heart weighing 450 grams [g]; normal weight is <400 g) • Arteriosclerotic CAD • Mild concentric left ventricular hypertrophy with wall thickness being 1.9 centimeters [cm]; normal thickness is 0.6-1.1 cm • A discrete 2 cm x 1 cm scar in the subendocardial and deep myocardium of the left ventricular wall (proximal medial posterior) suggesting a remote (old) heart attack (myocardial infarction [MI]) • No valvular disease • No sign of a pulmonary embolus (blood clot in the lung) • No evidence of smoke inhalation • Carboxyhemoglobin (COHb) level (measure of carbon monoxide in the blood) of 3.7% (normal for smokers)	⟲PDF

NIOSH REP. #	DATE OF INCIDENT	Title	CAUSE OF DEATH	AUTOPSY	AUTOPSY COMMENTS	PDF LINK
F2006-08	Nov 21, 2005	Firefighter/Emergency medical technician suffers an acute myocardial infarction and dies 3 days later—Pennsylvania	Per death certificate: "atherosclerotic heart disease" due to "acute myocardial infarction" as the cause of death. Perautopsy: "arterio-sclerotic cardiovascular disease (CVD) with acute myocardial infarction; pulmonary emboli" as the cause of death.	Yes	· Arteriosclerotic CVD · Extensive left ventricular MI · Cardiomegaly (enlarged heart: heart weighed 550 grams [g] · Biventricular hypertrophy · Microscopic examination revealed an "early thrombus at the site of the ventricular infarction" · Microscopic evidence of thromboemboli in the pulmonary arteries with pulmonary infarction	PDF
F2006-07	Feb 21, 2006	Two volunteer firefight-ers die when struck by exterior wall collapse at a commercial building fire overhaul—Alabama	Multiple blunt-force trauma	Yes	None	PDF
F2006-05	Nov 28, 2005	Career firefighter dies after engine he was driving collides with a tractor trailer—Alabama	Blunt-force injuries.	Not mentioned	None	PDF
F2006-04	Oct 24, 2005	Career lieutenant suffers sudden cardiac death at his station after making multiple runs during the day—Tennessee	Cardiac arrhythmia as the immediate cause of death due to hyper-tensive cardiovascular disease	Yes	· Cardiomegaly (enlarged heart) based on a weight of 700 grams · Left ventricular hypertrophy · Microscopic examination of the left ventricular myocar-dium (three sections) showed myocyte hypertrophy and small foci of subendocardial interstitial fibrosis (findings consistent with hypertensive heart disease) · Patent coronary arteries with mild atherosclerosis and no intracoronary thrombosis · Normal heart valves · No pulmonary embolus	PDF

NIOSH REP. #	DATE OF INCIDENT	Title	CAUSE OF DEATH	AUTOPSY	AUTOPSY COMMENTS	PDF LINK
F2006-03	Nov 05, 2005	Firefighter suffers sudden cardiac death during fire fighting operations—California	Atherosclerotic cardiovascular disease (CVD)	Yes	• Atherosclerotic CVD with complete narrowing (100%) of the right coronary artery and severe narrowing (75%) of the left main coronary artery and the left anterior descending coronary artery; no evidence of an intra-coronary blood clot (thrombus), but thrombolytic medications were given in the ED • Enlarged heart (cardiomegaly): heart weighed 450 grams (g) • No evidence of a premortem pulmonary thromboemboli (i.e., blood clots in the lungs) • Negative carboxyhemoglobin test • Negative drug and alcohol tests	⤓PDF
F2006-02	May 27, 2005	Lieutenant suffers sudden cardiac death after scba training—Florida	Atherosclerotic cardiovascular disease (CVD)	Yes	• Atherosclerotic CVD • Moderate cardiomegaly (enlarged heart: heart weighed 430 grams • Microscopic examination of the coronary arteries revealed an "apparent thrombus material identified in the lumen" with "evidence of re-canalization" (suggestive of a remote [old] myocardial infarction)) • Acute pulmonary edema • No evidence of thromboemboli in the pulmonary arteries • Negative carboxyhemoglobin test • Negative drug and alcohol tests	⤓PDF
F2006-01	May 13, 2004	Fire chief suffers sudden death during training—Alabama	Accidental multiple drug intoxication	Yes	• Enlarged heart (cardiomegaly): heart weighed 520 grams (g) • Mild atherosclerosis • Moderate to marked pulmonary edema • No evidence of thrombus in the coronary arteries • No evidence of thrombo-emboli in the pulmonary arteries • Positive blood tests for diazepam (Valium®), morphine (MS Contin®), venlafaxine (Effexor®) and methocarbamol (Robaxin®) • Positive urine test for gabapentin (Neurontin®) • No microscopic examinations were performed.	⤓PDF

NIOSH REP. #	DATE OF INCIDENT	Title	CAUSE OF DEATH	AUTOPSY	AUTOPSY COMMENTS	PDF LINK
F2005-35	Dec 02, 2005	Career captain dies and the driver/operator and a firefighter are severely injured in apparatus crash—Louisiana	Multiple blunt-force trauma and compressed asphyxiation	Yes	None	ᴸPDF
F2005-34	Nov 07, 2005	Career firefighter killed while riding manlift to assess a silo fire—Missouri	Positional asphyxia due to entrapment between the manlift and floor access opening	Not mentioned	None	ᴸPDF
F2005-33	Nov 04, 2005	Captain suffers pulmonary embolism during response to a medical call and later dies—New York	Pulmonary emboli (PE) due to probable deep vein thrombosis (DVT)	Yes	• Evidence of diffuse PE in both lungs Swollen right lower leg Mild hypertensive CVD: - No significant coronary artery atherosclerosis - Left ventricle walls slightly thickened/ hypertrophied (1.6 centimeters [cm]) (normal is 0.6 cm–1.1 cm)1 - Right ventricle walls slightly thickened/hypertrophied (0.6 cm) • Cardiomegaly (enlarged heart): heart weighed 540 grams (g) (normal is <400 g)3 • No microscopic examination of the heart muscle was performed • Negative drug screen	ᴸPDF

NIOSH REP. #	DATE OF INCIDENT	Title	CAUSE OF DEATH	AUTOPSY	AUTOPSY COMMENTS	PDF LINK

F2005-32	Mar 16, 2005	Sergeant suffers sudden cardiac death during training—Kentucky	Per death certificate: "acute myocardial infarction (MI)"
			
			 Per autopsy: "hypertensive/ischemic cardiovascular disease (CVD)" as the cause of death
				Yes	• Hypertensive/ischemic CVD
					 • Enlarged heart (cardiomegaly): heart weighed 500 grams (g) (normal is <400 g)1
					 • Biventricular hypertrophy
					   – Right ventricle measured 1 centimeter (cm) (normal is 0.3–0.5 cm)2
					   – Left ventricle measured 3 cm (normal is 0.76–0.88 cm)3 (normal echocardiographic measurement is 0.6–1.1 cm)4
					 • Mitral valve showed thickening of chordae tendineae
					 • No evidence of thrombus
					 • No evidence of thrombo-emboli
					 • Negative drug and alcohol tests
					 Microscopic examinations revealed the following:
					 • Diffuse myocardial fibrosis
					 • Old MI of lateral wall of left ventricular free wall, posterior aspect of left intraventricular septum
					 • Extensive old MI of left ventricular free wall
					 • Acute pulmonary congestion
						⤓PDF

F2005-30	May 31, 2005	Firefighter suffers sudden cardiac death during physical fitness training—New Jersey	"Calcific sclerosis of the aortic valve with aortic stenosis" as the cause of death, with "atherosclerotic coronary artery disease (CAD)" as a significant condition	Yes	• Calcific sclerosis of the aortic valve with aortic stenosis • Enlarged heart (cardiomegaly): heart weighed 580 grams (g) (normal is <400 g)1 • Atherosclerotic CAD • Severe narrowing (75%) of the left anterior descending coronary artery • No evidence of a premortem pulmonary thromboemboli Negative drug and alcohol tests	⤓PDF
F2005-29	Aug 27, 2005	Volunteer firefighter/rescue diver dies in training incident at a quarry—Pennsylvania	Drowning	Not mentioned	None	⤓PDF

NIOSH REP. #	DATE OF INCIDENT	Title	CAUSE OF DEATH	AUTOPSY	AUTOPSY COMMENTS	PDF LINK
F2005-27	Jun 21, 2005	Volunteer fire chief dies from injuries sustained during a tanker rollover—Utah	Blunt force injuries to the head and torso	Not mentioned	None	PDF
F2005-26	May 19, 2005	Recruit firefighter suffers heat stroke during physical fitness training and dies nine days later—Florida	Severe heat stroke with multisystem organ failure	No	None	PDF
F2005-25	Feb 23, 2005	Fire chief suffers sudden cardiac death at home after performing apparatus maintenance and conducting training—Texas	Severe three vessel atherosclerotic coronary artery disease (CAD)	Yes	• Atherosclerotic CAD: • Three vessel focal severe atherosclerotic CAD • Right ventricle chamber mildly dilated • Cardiomegaly (enlarged heart): heart weighed 460 grams (normal < 400 grams)1 • No pulmonary embolism • No evidence of a dissecting aortic aneurysm	PDF

NIOSH REP. #	DATE OF INCIDENT	Title	CAUSE OF DEATH	AUTOPSY	AUTOPSY COMMENTS	PDF LINK

F2005-24	Feb 06, 2005	Firefighter dies after responding to a call—New York	"Acute intoxication by the combined effects of propoxyphene (Darvon®) and cyclobenzaprine (Flexeril®)" as the cause of death and "hypertension" as another condition. The NIOSH investigator, like the medical examiner, concluded the FF died due to a drug intoxication, but cannot rule out the possibility of a cardiac arrhythmia associated with his hypertensive heart disease and subsequent left ventricular hypertrophy (LVH).	Yes	• Acute intoxication by the combined effects of propoxyphene and cyclobenzaprine CAD:
					  – Propoxyphene (Darvon®) blood level of 3.3 micrograms per milliliter (mcg/mL) and Norpropoxyphene blood level of 9.0 mcg/mL
					  – Cyclobenzaprine (Flexeril®) blood level of 90 nanograms per milliliter (ng/mL) and Norcyclobenzaprine with a positive blood level
					• Hypertensive cardiovascular disease:
					  – Left ventricular hypertrophy (LVH)(left ventricle wall thickness 1.5 centimeters [cm]) [normal 0.6cm–1.1cm])2
					  – Cardiomegaly (heart weight 500 grams [normal < 400 grams])3
					• No evidence of atherosclerotic coronary artery disease (CAD)
					• No evidence of pulmonary embolus
					Microscopic examination of the heart revealed myocyte hypertrophy (consistent with mild hypertensive cardiomyopathy) but no infarcts or myocarditis. The autopsy did not mention "myocytes in disarray," which is diagnostic of hypertrophic cardiomyopathy
						PDF

F2005-23	Mar 29, 2005	Firefighter/paramedic suffers a dissection of his aorta while participating in physical fitness training—Texas	"Aortic dissection and its sequelae" with "Hypertensive cardiovascular disease" being another significant condition	Yes	• Extent of the dissection: from proximal left and right coronary artery at the aortic root to the right common iliac artery • Cardiomegaly (a large heart) • Concentric left ventricular hypertrophy • No atherosclerotic coronary artery disease • Negative blood tests for illicit drugs or alcohol	PDF

NIOSH REP. #	DATE OF INCIDENT	Title	CAUSE OF DEATH	AUTOPSY	AUTOPSY COMMENTS	PDF LINK
F2005-22	Jan 09, 2005	Volunteer firefighter suffers cardiac death the morning after emergency medical technician training—North Carolina	Mitral valve failure	Yes	• Mitral valve prolapse—undulating leaflets with obvious hooding and thin, elongated chordae tendineae consistent with prolapse • Cardiomegaly—(530 grams, normal < 400 grams) with mild left ventricular hypertrophy – left ventricle thickness 1.4 cm (normal 0.6 to 1.1 cm), interventricular septum 1.6 cm (normal 0.6 to 1.1 cm) • Microscopic examination of the heart muscle showed widening of the individual myocytes and large, irregular, "boxcar" nuclei; focal increased interstitial fibrosis was present within the posterior left ventricle • Widely patent coronary arteries without evidence of significant atherosclerosis or thrombosis	⅊PDF

F2005-21	Feb 21, 2005	Wildland firefighter suffers sudden cardiac death after performing mop-up/overhaul operations at two wildland fires—Florida	Left anterior descending coronary artery thrombosis	Yes	• Attached thrombus (blood clot) with complete occlusion and focally a pinpoint lumen for about 1.5 cm in the proximal left anterior descending coronary artery
					• Chronic myocardial ischemia
					  – Fibrosis in the mid septum, just below the right aortic cusp, and near the apex of the heart (strongly suggestive of a remote heart attack)
					  – Heart weight 450 grams (normal < 400 grams)
					  – Left ventricular walls measure 1.4 centimeters (cm) in thickness (normal is 0.6 to 1.1 cm)
						⅊PDF

F2005-20	Jun 23, 2005	Firefighter/Emergency medical technician dies during the night at fire station—Arizona	"Oxycodone intoxication" as the cause of death with "mild thickening of the mitral valve" and "mild diffuse nephrosclerosis" as other conditions.	Yes	• Oxycodone intoxication (blood level of 0.85 milligrams per liter [mg/L])(therapeutic blood levels are less than 0.1 mg/L)2 • Mild thickening of the mitral valve	⅊PDF

NIOSH REP. #	DATE OF INCIDENT	Title	CAUSE OF DEATH	AUTOPSY	AUTOPSY COMMENTS	PDF LINK

F2005-19	Jun 19, 2004	Reserve firefighter suffers sudden cardiac death while working on a fuel reduction crew—Arizona	Atherosclerotic heart disease	Yes	• Atherosclerotic heart disease:
					  – 99% occlusion of left anterior descending coronary artery
					  – 99% occlusion of right coronary artery
					  – Left ventricular hypertrophy (wall thickness 1.4 centimeters [cm] [normal 0.6cm–1.1cm]) 1
					• Chronic pyelonephritis, left kidney
					• Negative drug and alcohol tests
						PDF

F2005-18	Aug 16, 2004	Airport firefighter suffers sudden cardiac death while on duty—South Carolina	Acute myocardial ischemia secondary to a cardiac arrhythmia of unknown etiology	Yes	• Enlarged heart (cardiomegaly) weighing 460 grams (normal < 400 grams)
					  – Left ventricular hypertrophy (wall thickness 1.3 centimeters [cm] [normal 0.6 cm–1.1 cm])
					  – Interventricular septum hypertrophy (wall thickness 1.5 cm [normal 0.6 cm–1.1 cm])
					• Right coronary artery bridging
					• Right ventricle mildly dilated
					• Microscopic findings
					  – Histological signs of ischemia
					  – Scattered myocytes indicative of hypertrophy
						PDF

F2005-17	Apr 20, 2005	Driver/operator dies due to a stroke while driving a fire engine to an alarm—Tennessee	"Acute subarachnoid hemorrhage" due to "rupture of sacular cerebral aneurysm" as the cause of death with "focal coronary artery atherosclerosis" as a significant condition	Yes	• Sacular aneurysm with rupture, basilar artery
					  – Acute subarachnoid hemorrhage, large amount, base of brain
					• Coronary artery atherosclerosis, focal and severe
					  – Severe narrowing (90%) of the left anterior descending coronary artery
					• Negative drug and alcohol tests
						PDF

NIOSH REP. #	DATE OF INCIDENT	Title	CAUSE OF DEATH	AUTOPSY	AUTOPSY COMMENTS	PDF LINK

F2005-16	Jan 21, 2004	Captain suffers an acute aortic dissection after responding to two alarms and subsequently dies due to hemopericardium—Pennsylvania	"Hemopericardium" due to an "aortic rupture" and "aortic dissection."	Yes	• Hemopericardium with 575 milliliters (mL) of primarily unclotted blood in pericardium
					• Rupture of the intrapericardial portion of the ascending thoracic aorta
					• Thoracoabdominal aortic dissection extending from the ascending aorta to the distal abdominal aorta (4 cm above the iliac bifurcation
					• History of hypertension
					  – Cardiomegaly, heart weighing 560 grams (normal < 400 grams)
					    • Moderate cardiac left ventricular hypertrophy
					• Minimal coronary artery disease (25% stenosis of the left anterior descending and right coronary arteries)
					• No evidence of infection in his left or right lung fields (pneumonia), although there were bilateral serosanguinous pleural effusions (200 mL on the left, 150 mL on the right)
					• Morbid obesity, with a body mass index of 51.9 kilograms/ meters2
						PDF

F2005-15	Apr 23, 2005	Career firefighter fatally injured in fall from apparatus—Texas	Blunt-force head injuries	Yes	None	PDF
F2005-14	Apr 22, 2004	Lieutenant suffers sudden cardiac death at the scene of a structure fire—South Carolina	"Acute myocardial infarction" due to "viral endocarditis"	No	Perform an autopsy on all onduty firefighter fatalities	PDF
F2005-13	Apr 18, 2005	A volunteer firefighter and volunteer assistant lieutenant die after a smoke explosion at a townhouse complex—Wyoming	Smoke inhalation and thermal burns to over 50% of their bodies	Not mentioned	None	PDF
F2005-12	Aug 23, 2004	Career firefighter/EMT dies in ambulance crash—Florida	Blunt-force head trauma	Not mentioned	None	PDF

NIOSH REP. #	DATE OF INCIDENT	Title	CAUSE OF DEATH	AUTOPSY	AUTOPSY COMMENTS	PDF LINK
F2005-11	Dec 13, 2004	Fire chief suffers sudden cardiac death while returning to the fire station after a structure fire—Georgia	"cardiorespiratory arrest" due to "ASCVD" (atherosclerotic cardiovascular disease)	No	Perform an autopsy on all onduty firefighter fatalities	PDF
F2005-10	Feb 15, 2005	Lieutenant suffers a heart attack while driving a squad truck and dies four days later—Georgia	"Anoxic encephalopathy" due to "sudden cardiac arrest" due to "atherosclerotic cardiovascular disease" as the cause of death	Yes	• Enlarged heart (cardiomegaly): heart weighed 450 grams • Ischemic heart disease: – Moderate narrowing (40%–50%) of the right coronary artery – Mild narrowing (30%–40%) of the left anterior descending coronary artery • No evidence of pulmonary thromboemboli • Negative drug and alcohol tests	PDF
F2005-09	Feb 19, 2005	Career fire captain dies when trapped by partial roof collapse in a vacant house fire – Texas	Smoke inhalation and thermal injuries	Not mentioned, but postmortem carboxyhemoglobin was reported at 26%	None	PDF
F2005-08	Oct 20, 2004	Fire chief suffers sudden cardiac death after responding to a motor vehicle crash—Texas	"arrhythmia" as the cause of death due to "coronary artery disease" (CAD) with "atrial fibrillation and sleep apnea" as other significant conditions	No	Perform an autopsy on all onduty firefighter fatalities	PDF
F2005-07	Feb 13, 2005	Career captain electrocuted at the scene of a residential structure fire—California	Electrocution	Not mentioned	None	PDF
F2005-06	Jan 07, 2005	Fire equipment operator suffers a heart attack at the scene of a medical call and dies in the hospital thirteen days later—South Carolina	"Myocardial infarction" due to "renal failure" due to "stroke"	No	Perform an autopsy on all onduty firefighter fatalities	PDF
F2005-05	Jan 20, 2005	Career captain dies after running out of air at a residential structure fire—Michigan	Smoke and soot inhalation with a carboxyhemoglobin level of 22.7%	Not mentioned	None	PDF

NIOSH REP. #	DATE OF INCIDENT	Title	CAUSE OF DEATH	AUTOPSY	AUTOPSY COMMENTS	PDF LINK
F2005-04	Jan 23, 2005	Career firefighter dies while exiting residential basement fire—New York	Smoke inhalation and burns of the head, torso and upper extremities	Yes	Note: Carboxyhemoglobin level was 24% saturation; third degree burns on approximately 63% of body surface area	⅄PDF
F2005-03	Jan 23, 2005	Career lieutenant and career firefighter die and four career firefighters are seriously injured during a three alarm apartment fire—New York	Mass trauma of the head, torso and extremities with multiple contusions of the extremities.	Not mentioned	None	⅄PDF
F2005-02	Dec 20, 2004	One probationary career firefighter dies and four career firefighters are injured at a two-alarm residential structure fire—Texas	Thermal injuries and smoke inhalation	Not mentioned	None	⅄PDF
F2005-01	Aug 14, 2004	Career firefighter dies after falling from tailboard and being backed over by engine—California	Multiple blunt-force injuries	Not mentioned	None	⅄PDF

NIOSH REP. #	DATE OF INCIDENT	Title	CAUSE OF DEATH	AUTOPSY	AUTOPSY COMMENTS	PDF LINK
F2004-46	Feb 23, 2004	Firefighter collapses and dies while assisting with fire suppression efforts at a residential fire—Ohio	"Arteriosclerotic and hypertensive heart disease" as the immediate cause of death with "morbid obesity" as a contributory condition	Yes	• Coronary artery disease (CAD) – Atherosclerotic occlusions of the native coronary arteries (remote) – Two vessel coronary artery bypass graft procedure (remote) – Occlusion of the lower coronary artery bypass graft (remote) – Open upper coronary artery bypass graft but with moderate calcific atherosclerosis – Myocardial infarct (heart attack) involving the anterior left ventricle and interventricular septum (remote) • Cardiomegaly (enlarged heart) weighing 780 grams (normal < 400 grams) • Mild narrowing of the arteries leading to the kidneys (arteriolonephrosclerosis), which is consistent with the diagnosis of hypertension • Morbid obesity; at 72 inches tall, the FF weighed 300 pounds for a body mass index (BMI) of 40.7. • A carboxyhemoglobin (COHb) measurement of <5% saturation. (Since the FF did not regain a heartbeat during resuscitation efforts, the 100% oxygen delivered via the endotracheal tube was unlikely to have significantly reduced the half-time of his COHb level. Thus, a level of <5% is unlikely to have contributed to his sudden death.) • Negative urine drug screen • Skull fractures	⅄PDF
F2004-45	Feb 04, 2004	Firefighter suffers sudden cardiac death after repacking a hose load on a fire engine—New Jersey	Arteriosclerotic cardio-vascular disease(CAD)	No	Perform an autopsy on all onduty firefighter fatalities	⅄PDF

NIOSH REP. #	DATE OF INCIDENT	Title	CAUSE OF DEATH	AUTOPSY	AUTOPSY COMMENTS	PDF LINK
F2004-43	Apr 27, 2004	One part-time firefighter dies and another is seriously injured when two fire engines collide at an intersection while responding to a fire—Illinois	Craniocerebral injuries with aspiration of blood within the lungs	Yes	None	PDF
F2004-42	Sep 26, 2004	Assistant chief suffers sudden cardiac death during response to boat fire—Wisconsin	Per death certificate, "severe arteriosclerotic cardiovascular disease" as the cause of death and "previous myocardial infarction" as other significant condition. Perautopsy: "severe coronary artery disease due to arteriosclerotic cardiovascular disease" as the cause of death	Yes	• Ischemic heart disease • Heavily calcified coronary arteries with severe atherosclerotic plaquing • Total occlusion of the mid right coronary artery • Significant narrowing (60%-95%) of the left anterior descending coronary artery • Mild plaquing in the left circumflex coronary artery • Cardiomegaly (enlarged heart) – heart weighing 450 grams (normal less than 400 grams) • A large area of white fibrosis with thinning of the posterior wall of the left ventricle • Severe dilatation of the heart chambers • No evidence of pulmonary thromboemboli • Negative drug and alcohol tests • Carboxyhemoglobin (carbon monoxide) test was not performed	PDF
F2004-41	Jun 08, 2004	Assistant chief suffers heart attack and dies after completing a walk test—Montana	"Sudden arrhythmia" as the immediate cause of death and "acute coronary artery thrombosis" as a contributing factor	Yes	• Enlarged heart (cardiomegaly): heart weighed 470 grams (normal < 400 grams) Ischemic heart disease: – Thrombus causing total occlusion in the right coronary artery – Significant narrowing (80% to 90%) of the left anterior descending coronary artery – Minor atherosclerosis in the circumflex artery • No evidence of pulmonary thromboemboli • Negative drug and alcohol tests	PDF

NIOSH REP. #	DATE OF INCIDENT	Title	CAUSE OF DEATH	AUTOPSY	AUTOPSY COMMENTS	PDF LINK
F2004-40	Sep 12, 2004	Career helitack firefighter dies in burnover during an initial attack at a wildland fire operation—California	Inhalation of products of combustion	Yes	None	⅄PDF
F2004-38	Mar 13, 2004	Probationary firefighter suffers sudden cardiac death during maze drill—Connecticut	Sudden death associated with rheumatic heart disease	Yes	Valvular heart disease: • Mitral valves thickened and fusion of the chordae tendineae" (strong cords of fibrous tissue fused together resulting in thickening and shortening of the mitral valve cusps) • Aortic valve fusion of the right and left cusps Normal heart size (380 grams), yet mild left atrial enlargement and dilatation • No significant coronary artery disease and no recent or remote evidence of a heart attack • Drug, alcohol, and carbon monoxide tests were negative	⅄PDF
F2004-37	Apr 08, 2004	Volunteer chief dies and two firefighters are injured by a collapsing church facade—Tennessee	Multiple blunt-force trauma and thermal injuries	Not mentioned	None	⅄PDF
F2004-36	Mar 11, 2004	Career firefighter drowns while conducting training dive—New Hampshire	Drowning	Not mentioned	None	⅄PDF

NIOSH REP. #	DATE OF INCIDENT	Title	CAUSE OF DEATH	AUTOPSY	AUTOPSY COMMENTS	PDF LINK

F2004-35	Jun 17, 2004	Firefighter suffers sudden cardiac death at his fire station—Georgia	"Cardiac dysrhythmia" as the immediate cause of death due to "massive cardiomegaly" due to "hypertensive heart disease."	Yes	• Hypertensive heart disease
					  – Severely enlarged heart (cardiomegaly) weighing 840 grams (normal < 400 grams), which is in the 95th percentile for body
					  – Markedly severely thickened left wall of the heart (concentric hypertrophy of the left ventricle)
					  – Moderate hardening of the kidney due to high blood pressure in the small arteries of the kidneys (arterionephrosclerosis)
					• Mild to moderate heart disease (atherosclerotic plaque blockage of three coronary arteries)
					  – Approximately 40%–50% blockages (stenosis) in each of the left anterior descending, circumflex, and right coronary arteries.
					• Mild thickening (myocytic hypertrophy) and mild to moderate scarring of the heart muscle (interstitial fibrosis), but no evidence of heart attack (acute or remote infarct)
					• Microscopic examination of the kidney showed kidney cancer (papillary renal cell carcinoma)
					His blood carboxyhemoglobin level (a test of carbon monoxide exposure) was not checked due to no exposure to fire smoke during his shift. No drug screen was performed
						PDF

F2004-32	May 13, 2004	Volunteer firefighter suffers cardiac arrest while battling a structure fire—New York	Acute myocardial infarction (heart attack) with physical exertion in a hot, humid environment listed as a contributing cause	No	Perform an autopsy on all on-duty firefighter fatalities	PDF
F2004-31	Aug 27, 2003	Volunteer firefighter suffers heart attack while battling structure fire and dies 6 days later—New York	Acute myocardial infarction	No	Perform an autopsy on all onduty firefighter fatalities	PDF

NIOSH REP. #	DATE OF INCIDENT	Title	CAUSE OF DEATH	AUTOPSY	AUTOPSY COMMENTS	PDF LINK
F2004-30	Mar 25, 2004	Firefighter collapses and suffers sudden cardiac death after responding to a vehicle fire—Kentucky	A cardiac event (arrhythmia) due to hypertensive and atherosclerotic cardio-vascular disease	Yes	• Cardiomegaly (an enlarged heart) weighing 560 grams (normal is less than 400 grams)1 • Dilated right ventricle • Scar of white fibrous tissue interlacing with brown myocar-dium (evidence of a remote [old] myocardial infarction [heart attack]) • Moderate to severe coronary artery atherosclerosis (left and right) • Possible thrombus in the right coronary artery	⬇PDF

F2004-28	May 31, 2002	Firefighter suffers sud-den cardiac death while performing work capacity test—California	Coronary atherosclerosis	Yes	• Atherosclerotic coronary artery disease (CAD)
					• Ischemic heart disease:
					  – Near complete occlusion of the most proximal branch of the left anterior descending coronary artery
					  – 90% stenosis in the left circumflex coronary artery
					  – 70% stenosis in the right coronary artery
					  – No superimposed acute thromboses or recent hemorrhages
					  – Enlarged heart (cardiomegaly) weighing 510 grams (normal < 400 grams)
					  – Left ventricular hypertrophy (wall thickness 1.7 centi-meters [cm] [normal 0.6cm-1.1cm])
					  – Interventricular septum hypertrophy (wall thickness 1.9cm [normal 0.6cm-1.1cm])
					• Mild to moderate perivascular fibrosis of the heart muscle on microscopic examination
					• No evidence of a pulmonary embolus
					• Vitreous (eye) chemistries showed elevations in sodium and chlorine consistent with dehydration
					• Negative drug and alcohol tests
						⬇PDF

NIOSH REP. #	DATE OF INCIDENT	Title	CAUSE OF DEATH	AUTOPSY	AUTOPSY COMMENTS	PDF LINK

F2004-26	Feb 01, 2003	Firefighter-engineer suffers sudden cardiac death while performing strenuous fire station maintenance—California	"Atherosclerotic coronary artery disease" as the immediate cause of death with "hypertension"	Yes	• Cardiomegaly (an enlarged heart) weighing 510 grams (upper limit of normal is 400 grams)
					• Moderate biventricular dilatation (LV—4 centimeters (cm) internal diameter; RV- 4.0 X 4.5 cm internal diameter)
					• Atherosclerotic coronary artery disease
					  – 75% stenosis of the right coronary artery with a right dominant coronary artery system
					  – 75% stenosis in the left main artery
					  – 90% stenosis in the left anterior descending artery
					  – 90% stenosis in the left circumflex artery
					  – No coronary artery thrombus (blood clot)
					• No pulmonary emboli (blood clot in the lung vasculature)
					• Negative drug screen
						⤓PDF

F2004-25	Mar 22, 2003	Volunteer firefighter suffers sudden cardiac death during fire suppression at a structural fire—Indiana	Acute, occlusive thrombosis of the left circumflex coronary artery secondary to coronary artery atherosclerosis	Yes	• Moderate to severe coronary atherosclerosis as detailed below.
					  – Acute occlusive thrombus present within the mid to distal portion of the left circumflex coronary artery
					  – 80%-95% stenosis of the distal left anterior descending coronary artery and the right posterior descending coronary artery
					  – Intraplaque hemorrhage at the mid portion of the left anterior descending coronary artery
					  – 40%-60% stenosis of the mid portions of left anterior, left circumflex, and right coronary arteries
					  – Cardiomegaly (enlarged heart weighing 550 grams, normal is less than 400 grams)
					  – Extensive biventricular dilatation
					  – Concentric left ventricular hypertrophy (LVH)
					• Negative drug and alcohol test results
					• Carboxyhemoglobin level (a measure of carbon monoxide exposure) of 2.2% (10 hours after the initiation of oxygen). The normal lab value for moderate to heavy smokers is 4-15%. The FF did have a history of smoking.
						⤓PDF

NIOSH REP. #	DATE OF INCIDENT	Title	CAUSE OF DEATH	AUTOPSY	AUTOPSY COMMENTS	PDF LINK
F2004-24	Apr 15, 2004	Acting fire chief suffers heart attack after shift and dies—Alaska	Cardiac event	No	Perform an autopsy on all onduty firefighter fatalities	⬇PDF
F2004-23	May 06, 2002	Firefighter suffers unwitnessed sudden cardiac death after responding to mobile home fire—South Carolina	"Cardiac arrhythmia" as the immediate cause of death due to "stress of fighting fire" and "cardiomegaly with dilatation" as contributing factors	Yes	• Cardiomegaly (enlarged heart)—weighing 576 grams (normal less than 400 grams) • Mild atherosclerotic cardiovascular disease – 50% occlusion in the right coronary artery – 50% occlusion in the left anterior descending artery • Biventricular hypertrophy with dilatation (left ventricle thickness of 1.5 cm at the anterior papillary muscle) • Biventricular hypertrophy with dilatation (left ventricle thickness of 1.5 cm at the anterior papillary muscle) • No evidence of thrombi, emboli, or fibrosis on gross pathology • Drug and alcohol tests were negative • Carboxyhemoglobin (carbon monoxide) test was negative (less than 10%) • Hepatosplenomegaly	⬇PDF

NIOSH REP. #	DATE OF INCIDENT	Title	CAUSE OF DEATH	AUTOPSY	AUTOPSY COMMENTS	PDF LINK

F2004-22	Apr 10, 2004	Career firefighter/Emergency medical technician suffers sudden death 5 hours after participating in emergency response—South Carolina	Per death certificate: "atherosclerotic coronary artery disease" Per autopsy: "acute cocaine intoxication" as the immediate cause of death with hypertension and atherosclerotic coronary heart disease as contributing factors.	Yes	• Extensive atherosclerotic coronary artery disease with near total occlusion of three coronary arteries (left anterior descending, first diagonal, and right coronary artery
					• Acute thrombus ("acute organizing occlusive fibrin thrombi")
					• Old myocardial infarction of the posterior left ventricle
					• Cardiomyopathy
					  – Cardiomegaly (an enlarged heart) weighing 680 grams (upper limit of normal is 400 grams
					  – Hypertensive cardiovascular disease
					  – Biventricular dilatation
					• Acute Cocaine Intoxication
					  – Cocaine, Blood: 0.48 mg/L
					  – Ecgonine methyl ester, blood: 0.67 mg/L
					  – Benzoylecgonine, blood: 0/76 mg/L
					  – Benzoylecgonine, urine: 4.9 mg/L
						PDF

NIOSH REP. #	DATE OF INCIDENT	Title	CAUSE OF DEATH	AUTOPSY	AUTOPSY COMMENTS	PDF LINK
F2004-21	Apr 18, 2004	District chief suffers sudden cardiac death at home after experiencing symptoms consistent with heart disease at his station—Illinois	"Cardiac arrhythmia" as the cause of death due to "ischemic heart disease."	Yes	• Marked atherosclerosis with – 60% narrowing of the right coronary artery – 60% narrowing of the left main coronary artery – 90% narrowing of the left anterior descending coronary artery – 90% narrowing of the circumflex coronary artery • Recent thrombus in the circumflex coronary artery, resulting in total occlusion • No evidence of remote myocardial infarction or myocardial fibrosis • Mild left ventricle hypertrophy with: – Left ventricle thickness 1.6 cm (normal is 0.76–0.88 cm) – Interventricular septum thickness 1.5 cm (normal echographic measurement is 0.6–1.1 cm) • Cardiomegaly (enlarged heart) weighing 420 grams (normal is less than 400 grams)	▶PDF
F2004-20	Dec 25, 2003	Assistant chief dies after suffering aortic dissection during a fire alarm response—Connecticut	"Multi-organ system dysfunction" due to "cardiac arrest" as the immediate cause of death with "aortic dissection" as a significant condition	No	Perform an autopsy on all onduty firefighter fatalities	▶PDF
F2004-19	Apr 30, 2004	Career firefighter dies from injuries sustained in fall from apparatus—Massachusetts	Blunt-force head trauma	Not mentioned	None	▶PDF

NIOSH REP. #	DATE OF INCIDENT	Title	CAUSE OF DEATH	AUTOPSY	AUTOPSY COMMENTS	PDF LINK
F2004-18	Jan 14, 2004	Volunteer firefighter suffers sudden cardiac death after participating in emergency responses—Maryland	Acute myocardial infarction	Yes	• Cardiomegaly (an enlarged heart) weighing 670 grams (upper limit of normal is 400 grams) • Four chamber dilatation • Biventricular hypertrophy • Atherosclerotic CAD • Acute plaque rupture and thrombosis, left obtuse marginal artery (recent heart attack) • Total occlusion of mid left circumflex artery by healed plaque rupture with organized and recanalized thrombus (old, healed heart attack) • Healed transmural infarction, posterolateral left ventricle at base • Diffuse 50%-90% narrowing of coronary arteries • Penetrating organized thrombus, right atrium; undetermined etiology	⅄PDF
F2004-17	Mar 13, 2004	Career battalion chief and career master firefighter die and 29 career firefighters are injured during a five alarm church fire—Pennsylvania	Victim #1: asphyxiation due to compression of the body by building debris; blunt-force trauma of the head, neck, pelvis, and extremities were contributory causes of death. Victim #2: asphyxiation due to compression of the body by building debris with blunt force trauma of the head and extremities.	Not mentioned	None	⅄PDF

NIOSH REP. #	DATE OF INCIDENT	Title	CAUSE OF DEATH	AUTOPSY	AUTOPSY COMMENTS	PDF LINK
F2004-16	Dec 18, 2003	Firefighter dies at home after shift—Maryland	Arteriosclerotic cardio-vascular disease	Yes	The carboxyhemoglobin level (a measure of exposure to carbon monoxide) was not measured • Severe arteriosclerotic cardiovascular disease • Old subendocardial infarct of left ventricle • Biventricular dilatation • Cardiomegaly (heart weighing 550 grams with normal less than 400 grams4) • Aortic atherosclerosis • Cardiac valves were unremarkable • No thrombi were found • Drug and alcohol tests were negative	PDF
F2004-15	Mar 03, 2004	Forest ranger/Firefighter drowned after catastrophic blow-out of right front tire—Florida	Drowning with no evidence of trauma-related injuries	Not mentioned	Postcrash blood alcohol content (BAC) and drug screening tests were negative	PDF
F2004-14	Apr 04, 2004	Career firefighter dies and two career captains are injured while fighting night club arson fire—Texas	High thermal exposure	Yes	None	PDF
F2004-13	Apr 18, 2003	Firefighters suffers fatal pulmonary embolism after knee surgery for a work-related injury—North Carolina	Massive pulmonary embolism (PE) due to a deep vein thrombus (DVT) due to knee injury that was treated surgically	Yes	• Acute pulmonary embolus • Remote pulmonary emboli (at least two weeks old) • Deep vein thrombosis in the right leg • Evidence of hypertensive cardiovascular disease • Severe focal atherosclerotic coronary artery disease • No ethanol (alcohol) or salicylates (aspirin) detected	PDF
F2004-12	Jan 27, 2004	Firefighter/Paramedic dies after performing physical fitness training—Florida	Aortic valve stenosis	Yes	• Aortic valve stenosis • Cardiomegaly (heart weighing 440 grams with normal less than 400 grams) • Mild, patchy, interstitial fibrosis [as determined by micro-scopic examination] • No evidence of thromboemboli • No evidence of atherosclerosis	PDF

NIOSH REP. #	DATE OF INCIDENT	Title	CAUSE OF DEATH	AUTOPSY	AUTOPSY COMMENTS	PDF LINK
F2004-11	Feb 13, 2004	Career lieutenant killed and firefighter injured by gunfire while responding to medical assistance call—Kentucky	Multiple gunshot wounds	Not mentioned	None	PDF
F2004-10	Feb 18, 2004	Career firefighter dies searching for fire in a restaurant/lounge—Missouri	Smoke inhalation	Not mentioned	An independent toxicology report listed the victim's carbon monoxide level at 51% saturation. There was no notable trauma	PDF
F2004-09	Apr 02, 2001	Firefighter/Driver/Engineer suffers heart attack and dies at the end of his 24-hour shift—Hawaii	Acute myocardial infarction and coronary artery thrombosis due to atherosclerotic cardiovascular disease	Yes	• Blood clot (thrombus) in one of the coronary arteries (mid-left anterior descending artery) • Moderate to severe coronary atherosclerosis • A large heart weighing 470 grams (normal < 400 grams) • Thickened left wall of the heart (left ventricular hypertrophy) • A negative drug screen	PDF
F2004-08	Jan 21, 2004	Firefighter suffers sudden cardiac death after emergency recall—Massachusetts	"Acute sudden cardiac death syndrome" due to "acute myocardial infarction" as the immediate cause of death and "obesity" as another significant condition	No	Perform an autopsy on all onduty firefighter fatalities	PDF
F2004-07	Nov 18, 2002	Fire chief dies after performing service call—Connecticut	"ASCVD" (atherosclerotic coronary vascular disease) as the immediate cause of death, with hyperlipidemia and smoking as contributing factors	Yes	• Arteriosclerosis, with 95% occlusion of the left main coronary artery and 90% occlusion of the right coronary artery • No CO level taken	PDF

NIOSH REP. #	DATE OF INCIDENT	Title	CAUSE OF DEATH	AUTOPSY	AUTOPSY COMMENTS	PDF LINK
F2004-06	Nov 17, 2003	Firefighter/Paramedic suffers sudden cardiac death while performing physical fitness training—Washington	Occlusive atherosclerotic cardiovascular disease	Yes	• Heart weighing 400 grams • Atherosclerotic cardiovascular disease • 0.7 centimeter area of increased consistency which is slightly grayer than adjacent areas in the posterior lateral aspect of the left ventricular wall in the apical third • No thrombi or emboli • No fibrosis • Drug and alcohol tests were negative	⅄PDF
F2004-05	Jan 09, 2004	Residential basement fire claims the life of career lieutenant—Pennsylvania	Smoke and soot inhalation and thermal burns	Not mentioned	None	⅄PDF
F2004-04	Dec 16, 2003	Career firefighter dies of carbon monoxide poisoning after becoming lost while searching for the seat of a fire in warehouse—New York	Smoke inhalation with a carboxyhemoglobin (cohb) level of 74.8% in the emergency department	Yes	None	⅄PDF
F2004-03	Nov 17, 2003	Career captain/safety officer dies in a single motor vehicle crash while responding to a call—Kansas	Probable positional asphyxia	Not mentioned	None	⅄PDF
F2004-02	Nov 29, 2003	Basement fire claims the life of volunteer firefighter – Massachusetts	Smoke and soot inhalation	Not mentioned	None	⅄PDF
F2004-01	Oct 24, 2003	District chief dies after suffering a heart attack—Texas	Myocardial infarction	No	Perform an autopsy on all onduty firefighter fatalities	⅄PDF
F2003-41	Oct 17, 2003	Live-fire exercise in mobile flashover training simulator injures five career firefighters—Maine	No deaths		None	⅄PDF
F2003-40	Feb 26, 2003	Airport firefighter suffers sudden cardiac death at fire station—ArKansas	Myocardial infarction	No	Perform an autopsy on all onduty firefighter fatalities	⅄PDF
F2003-39	Jul 21, 2003	Firefighter suffers sudden cardiac death in parking lot of fire station—Tennessee	Myocardial infarction	No	Perform an autopsy on all onduty firefighter fatalities	⅄PDF

NIOSH REP. #	DATE OF INCIDENT	Title	CAUSE OF DEATH	AUTOPSY	AUTOPSY COMMENTS	PDF LINK
F2003-38	Oct 07, 2003	Firefighter dies after performing ventilation at a fire in a two-story dwelling—Pennsylvania	"Ischemic heart disease" as the immediate cause of death and "smoke inhalation" as a significant condition	Yes	• Cardiomegaly (heart weighing 552 grams, with normal less than 400 grams) • Mild atherosclerotic coronary artery disease involving the left main coronary artery • Remote (old) myocardial infarction involving the apex and left posterior ventricular wall • Fibrosis within the apex and left posterior ventricular wall (as determined by microscopic examination) • Evidence of smoke inhalation (moderate amount of soot in the trachea and large airways of both lungs) • Carboxyhemoglobin (measure of carbon monoxide in the blood) level negative	PDF
F2003-37	Oct 27, 2003	Volunteer assistant chief is struck and killed at road construction site—Minnesota	Craniocerebral injuries and closed head trauma	Yes	None	PDF
F2003-36	Oct 29, 2003	A career firefighter was killed and a career captain was severely injured during a wildland/urban interface operation—California	Thermal injuries	Yes	Extensive burns over the entire body, no evidence of underlying cardiovascular or pulmonary disease, and a carboxyhemoglobin level of 27% (confirming significant exposure to carbon monoxide prior to his death)	PDF
F2003-35	Sep 27, 2003	Firefighter suffers a heart attack after responding to a rubbish fire at a two-story apartment building—New York	"Acute myocardial infarction" as the immediate cause of death and "diabetes mellitus" and "hypertension" as other significant conditions	No	None	PDF
F2003-34	Jul 10, 2003	Volunteer firefighter/fire service products salesman dies after being struck by dislodged rescue airbag—South Dakota	Closed head wound	Not mentioned	None	PDF

NIOSH REP. #	DATE OF INCIDENT	Title	CAUSE OF DEATH	AUTOPSY	AUTOPSY COMMENTS	PDF LINK
F2003-33	Aug 06, 2003	Career firefighter/emergency medical technician dies and paramedic is injured in a three-vehicle collision—Nebraska	Blunt-force trauma due to a motor vehicle accident	Not mentioned	None	PDF
F2003-32	Oct 01, 2003	Two firefighters die and eight firefighters are injured from a silo explosion at a lumber company—Ohio	Multiple blunt-force injuries	Not mentioned	None	PDF
F2003-31	Apr 14, 2003	Lieutenant suffers sudden cardiac death after performing forcible entry requiring heavy physical exertion—Georgia	"Cardiac dysrhythmia" due to "atherosclerotic coronary artery disease" as the immediate cause of death and "superimposed physical exertion" as a contributing factor	Yes	• Cardiomegaly (heart weighing 530 grams, with normal less than 400 grams1) • Atherosclerotic coronary artery disease • Left ventricular hypertrophy (15 millimeters (mm) thick; normal between 7.6-8.8 mm • Interventricular septal hypertrophy (16mm thick; normal is 6 to 11 mm) • No thromboemboli are recovered from the main, right, or left pulmonary arteries or their segmental branches • No obvious soot in the nares or oral cavity • Microscopic sections of the right ventricle, left ventricle, and interventricular septum do not reveal significant myocardial inflammation, infarct, hemorrhage, fibrosis, or neoplasia • Drug and alcohol tests were negative	PDF
F2003-30	Jul 28, 2003	One volunteer lieutenant dies and a volunteer firefighter is seriously injured in a motor vehicle rollover incident while enroute to a trailer fire—North Carolina	Acute intracranial injuries	Not mentioned	None	PDF

NIOSH REP. #	DATE OF INCIDENT	Title	CAUSE OF DEATH	AUTOPSY	AUTOPSY COMMENTS	PDF LINK
F2003-29	Dec 15, 2002	Firefighter suffers heart attack at the scene of a structure fire and dies 2 months later—Indiana	"Acute myocardial infarction" due to "atherosclerotic cardio-vascular disease" as the immediate cause of death and "chronic obstructive pulmonary disease (COPD)" as a contributing factor	No	Perform an autopsy on all onduty firefighter fatalities	PDF
F2003-28	Aug 08, 2003	Live-fire training exercise claims the life of one recruit firefighter and injures four others—Florida	Atherosclerotic cardio-vascular disease" due to "COPD"	No	Perform an autopsy on all onduty firefighter fatalities	PDF
F2003-27	Jan 25, 2003	Fire captain suffers sudden cardiac death during a live-fire training exercise—North Carolina	"Acute myocardial infarction" (heart attack) as the cause of death due to "coronary atherosclerosis" with a "prior myocardial infarction" being a significant contributing factor	Yes	Severe atherosclerotic coronary artery disease • 100% blockage of the right coronary artery, the circumflex coronary artery, and the first diagonal coronary artery in the area of the stent • A seven centimeter scar (due to his heart attack in 1999) in the posterior and lateral left ventricular wall • A two centimeter hyperemic area near the first diagonal coronary artery which probably represented the early signs of a recent (acute) heart attack (myocardioal infarction) Carboxyhemoglobin level was less than 5% suggesting carbon monoxide exposure was not a significant factor in the Captain's sudden death	PDF

NIOSH REP. #	DATE OF INCIDENT	Title	CAUSE OF DEATH	AUTOPSY	AUTOPSY COMMENTS	PDF LINK
F2003-26	Feb 20, 2003	Firefighter suffers sudden cardiac death at his fire station—Oregon	Per death certificate: "ischemic heart disease" as the immediate cause of death due to "atherosclerotic coronary heart disease." Per autopsy: "arteriosclerotic cardiovascular disease" as the cause of death	Yes	• Moderate calcification and atherosclerotic narrowing of the coronary arteries • Softening and dark discoloration of the left ventricle, more toward the apex" [a finding suggestive of a recent myocardial infarct (MI) (otherwise known as a heart attack)] • No scars suggestive of old/remote heart attacks • No evidence of a blood clot (embolus) in the pulmonary arteries • Microscopic examination of the heart muscle showed no inflammation, necrosis, or scarring • Blood carboxyhemoglobin level (a test of carbon monoxide exposure) was not checked due to no exposure to fire smoke during his shift, and no drug screen was performed	PDF
F2003-25	May 14, 2003	Career Federal firefighter dies from injuries sustained at prescribed burn—Arizona	Adult Respiratory Distress Syndrome secondary to severe inhalation injury with cardiovascular compromise.	None mentioned	Note: Thirty-six % of the victim's body surface area had second- and third-degree burn injuries and he had a significant inhalation injury	PDF
F2003-24	Jan 21, 2003	Firefighter suffers fatal heart attack while performing physical fitness training—Missouri	Per death certificate: "acute myocardial infarction" due to "atherosclerotic coronary artery disease" Per autopsy: "coronary atherosclerosis" followed by "stenosis, thrombosis, chronic myocardial infarct, subacute myocardial infarcts, acute myocardial infarct, and acute ischemic change" as the cause(s) of death.	Yes	• A large heart (435 grams with normal less than 400 grams) • Significant coronary atherosclerosis • Stent placement in three coronary arteries [left anterior descending (LAD), left circumflex, right coronary artery] • Evidence of old heart attacks (myocardial infarcts or MIs) • Subacute MIs • Acute (recent) MI in the interventricular septum and left ventricular free wall Since the captain was not involved in any fire suppression duties that day, a carboxyhemoglobin level (a measure of exposure to carbon monoxide) was not checked	PDF

NIOSH REP. #	DATE OF INCIDENT	Title	CAUSE OF DEATH	AUTOPSY	AUTOPSY COMMENTS	PDF LINK
F2003-23	Jun 26, 2003	Volunteer assistant chief dies in tanker rollover—New Mexico	Multiple injuries consistent with a rollover motor vehicle crash	Not mentioned	None	⤵PDF
F2003-22	Dec 13, 2002	Volunteer firefighter suffers sudden cardiac death after completing emergency medical technician (emt) written examination—Texas	Acute myocardial infarction	No	Perform autopsies on all on-duty firefighter fatalities	⤵PDF
F2003-21	Feb 12, 2003	Firefighter recruit suffers sudden cardiac death during physical ability training—Texas	Cardiac hypertrophy, "biventricular dilatation and cardiomegaly"	Yes	• Cardiac hypertrophy • Biventricular and right atrial dilatation • Cardiomegaly (an enlarged heart) weighing 440 grams • The coronary arteries are free of atherosclerosis • The cardiac valves are unremarkable • No blood clots in the pulmonary vessels, therefore no evidence of a pulmonary embolus	⤵PDF
F2003-20	May 22, 2003	Junior volunteer firefighter is killed while responding to a brush fire with an intoxicated driver—Wyoming	Massive trauma	Not mentioned	None	⤵PDF
F2003-19	Jun 16, 2003	Volunteer firefighter killed after his privately owned vehicle hydroplaned and struck a billboard signpost—Kentucky	Massive head and chest trauma	Not mentioned	None	⤵PDF
F2003-18	Jun 15, 2003	Partial roof collapse in commercial structure fire claims the lives of two career firefighters—Tennessee	Victim #1: thermal burns. Victim #2: thermal inhalation injury	Not mentioned	None	⤵PDF
F2003-17	May 18, 2003	Volunteer training/Safety officer dies from injuries received in fall from pick-up truck following training exercise—Tennessee	Blunt-force injury to the head	Not mentioned	None	⤵PDF

NIOSH REP. #	DATE OF INCIDENT	Title	CAUSE OF DEATH	AUTOPSY	AUTOPSY COMMENTS	PDF LINK
F2003-16	Feb 23, 2003	Volunteer fire police captain dies from injury-related complications after being struck by motor vehicle while directing traffic—New Jersey	Severe pneumonia as a consequence of complications due to a pedestrian motor vehicle accident	Not mentioned	None	PDF
F2003-15	Apr 03, 2003	Volunteer firefighter dies in tanker rollover—Ohio	Compressive asphyxia	Not mentioned	None	PDF
F2003-14	Mar 19, 2003	Volunteer captain killed in fire apparatus crash while responding to a training exercise—Oregon	Positional asphyxiation	Not mentioned	None	PDF
F2003-13	Mar 18, 2003	Volunteer firefighter killed while walking across an interstate highway responding to a motor vehicle incident—Texas	Severe craniocerebal injuries	Yes	None	PDF
F2003-12	Mar 31, 2003	Career firefighter dies and two career firefighters injured in a flashover during a house fire—Ohio	Severe third degree burns	Not mentioned	None	PDF
F2003-11	Feb 07, 2001	Firefighter collapses and dies at the scene of residential fire—Florida	Hypertensive and arteriosclerotic heart disease	Yes	• Enlarged heart (concentric left ventricular hypertrophy) • Coronary atherosclerosis • Four-vessel bypass • Pulmonary edema • Cerebral edema His blood carboxyhemoglobin level was not checked	PDF

NIOSH REP. #	DATE OF INCIDENT	Title	CAUSE OF DEATH	AUTOPSY	AUTOPSY COMMENTS	PDF LINK
F2003-10	May 22, 2002	Firefighter suffers sudden cardiac death during a medical emergency response—California	Complication of hypertrophic cardiomyopathy	Yes	• An enlarged heart weighing 565 grams (normal less than 400 grams) • Thickened left ventricle of 2.0 centimeters (cm) in diameter (normal <1.3) • Thickened right ventricle of 1.0 centimeters (cm) in diameter (normal <0.8) • Minimal atherosclerosis in the coronary arteries • surgical patch and scar at the apex of the heart's left ventricle consistent with a well healed surgical repair of an old knife wound to the chest • Bilateral fibrous pleural adhesions and pericardial sac adhesions also consistent with his old knife wound to the chest • No evidence of a blood clot (embolus) in the pulmonary arteries • A negative drug screen of illicit drugs (e.g. cocaine) • Histology (microscopic) examination of the heart tissue was not performed	PDF

NIOSH REP. #	DATE OF INCIDENT	Title	CAUSE OF DEATH	AUTOPSY	AUTOPSY COMMENTS	PDF LINK
F2003-09	Jan 14, 2003	Firefighter suffers fatal heart arrhythmia at structure fire—Illinois	"Cardiac arrhythmia" due to "heart disease" as the immediate cause of death and "exertion while fighting a house fire" as a contributing factor	Yes	• Atherosclerotic cardiovascular disease • Coronary artery atherosclerosis • Remote myocardial infarct (heart attack), posterior/inferior wall of left ventricle with moderate thinning of the wall • Diffuse mild and focal moderate aortic atherosclerosis • Clinical history of hypercholesterolemia • Hypertensive cardiovascular disease • Cardiac hypertrophy (490 grams) • Clinical history of hypertension • Tobacco pneumonitis with early pulmonary emphysema and chronic bronchitis • History of cigarette smoking • No ethanol or illicit drugs were detected Carboxyhemoglobin level was 2.2%, indicating the captain had inhaled some but not an excessive amount of carbon monoxide; possibly due to his cigarette smoking	⅄ PDF
F2003-08	Nov 12, 2002	Sudden cardiac death due to myocardial sarcoidosis claims the life of an onduty firefighter—Connecticut	Myocardial sarcoidosis	Yes	• Myocardial sarcoidosis (noncaseating granulomas) of the heart, lung, lymph nodes, and liver • The right heart ventricle revealed lymphocytic infiltrates and noncaseating granulomas that are confluent, with dense fibrosis between the granulomas. There is also some nodule involvement of the papillary muscles and interventricular septum (all findings consistent with sarcoidosis involving the heart) • Cardiomegaly (an enlarged heart) weighing 640 grams • No atherosclerotic disease of the coronary arteries	⅄ PDF
F2003-07	Jan 13, 2003	Career firefighter/emergency medical technician dies from injuries sustained in fall from apparatus—California	Multiple traumatic injuries	Not mentioned	None	⅄ PDF

NIOSH REP. #	DATE OF INCIDENT	Title	CAUSE OF DEATH	AUTOPSY	AUTOPSY COMMENTS	PDF LINK
F2003-06	Sep 19, 2002	Firefighter dies from progressive respiratory failure—Massachusetts	Progressive respiratory failure and clinical history of adult respiratory distress syndrome due to inhalational injuries		• Pulmonary congestion • Tracheobronchial tree is diffusely obstructed by mucoid-type material • Areas of squamous metaplasia in the midtrachea • Chronic inflammatory infiltrate within the mucosa and submucosa of the lungs • Diffuse alveolar damage in the reparative phase • Interstitial fibrosis of the lungs • No significant narrowing of the coronary arteries • Cardiomegaly (enlarged heart) (466 grams)	PDF
F2003-05	Jan 19, 2003	Career firefighter/emergency medical technician dies in ambulance crash—Texas	Blunt-force injuries	Not mentioned	None	PDF
F2003-04	Jan 20, 2003	Career firefighter dies from injuries received during a chimney and structural collapse after a house fire—Pennsylvania	Compressional asphyxia as a result of being trapped from falling debris	Not mentioned	None	PDF
F2003-03	Jan 19, 2003	Volunteer firefighter dies following nitrous oxide cylinder explosion while fighting a commercial structure fire—Texas	Thermal injuries, with smoke inhalation and blast effect	Yes	Preliminary autopsy findings indicated that he had received significant blast injuries, i.e., both eardrums were ruptured and there was concussive damage to his lungs	PDF
F2003-02	Dec 05, 2002	Firefighter suffers fatal heart attack at two-alarm structure fire—Texas	Atherosclerotic and hypertensive cardio-vascular disease	Yes	• Severe occlusive coronary artery disease • Remote infarct of the posterior wall of the left ventricle and the posterior aspect of the interventricular septum • Cardiomegaly (an enlarged heart) weighing 560 grams (normal is less than 400 grams) • Concentric left ventricular hypertrophy (free wall width 1.5 centimeters thick) Carboxyhemoglobin level was less than one percent, indicating inhaled carbon monoxide was not a factor in his death	PDF

NIOSH REP. #	DATE OF INCIDENT	Title	CAUSE OF DEATH	AUTOPSY	AUTOPSY COMMENTS	PDF LINK
F2003-01	May 29, 2002	Firefighter dies during night at fire station—Mississippi	"Sudden cardiac death" secondary to hypertensive heart disease and coronary artery disease	Yes	• Coronary artery disease (CAD) • An enlarged heart (460 grams). • Histologic and visual inspection of the heart failed to reveal evidence of acute or remotein farction • Microscopic analysis reveals ischemic change manifested by nuclear hyperchromatism, nuclear outline irregularity, and nuclear enlargement of the nuclei of the individual rhabdomycytes	PDF
F2002-50	Nov 25, 2002	Structural collapse at an auto parts store fire claims the lives of one career lieutenant and two volunteer firefighters—Oregon	Asphyxiation	Not mentioned	None	PDF
F2002-49	Nov 01, 2002	Volunteer lieutenant dies following structure collapse at residential house fire—Pennsylvania	Traumatic compressional asphyxia	Not mentioned	None	PDF

NIOSH REP. #	DATE OF INCIDENT	Title	CAUSE OF DEATH	AUTOPSY	AUTOPSY COMMENTS	PDF LINK
F2002-48	Aug 28, 2001	Firefighter suffers sudden cardiac death at a structural fire—New York	"Hypertrophic cardio-myopathy with myocar-dial arteriolarsclerosis" as the immediate cause of death with a "myxomatous mitral value"	Yes	• Widely patent (open) coronary arteries • An enlarged heart weighing 440 grams • Slightly thickened left ventricle of 1.4 centimeters (cm) in diameter • Microscopic changes of the heart muscle • Myocyte bundles arranged in a sinusoidal and other irregular patterns • Focal myocyte hypertrophy with occasional multipolarity • Widened interstitium with fibrosis in irregular patterns • Arterioles with thickened intima • No inflammatory infiltrates, contraction banding, or hypereosinophilia • Marked billowing of both leaflets of the mitral valve • No evidence of a blood clot (embolus) in the pulmonary arteries • Blood carboxyhemoglobin level was less than 3%, sug-gesting the victim was not exposed to excessive carbon monoxide levels	PDF

NIOSH REP. #	DATE OF INCIDENT	Title	CAUSE OF DEATH	AUTOPSY	AUTOPSY COMMENTS	PDF LINK
F2002-47	Jan 13, 2001	Firefighter suffers a heart attack and dies after performing "ventilation-entry-search" activities in a five-story apartment building fire—New York	"Hypertensive and arteriosclerotic heart disease" as the immediate cause of death and "smoke inhalation" as an other significant condition	Yes	• Hypertensive and arteriosclerotic cardiovascular disease • 60% calcific atherosclerotic stenosis in left main stem coronary artery • 30% stenosis in proximal right coronary artery • Valves have focal atherosclerosis • Cardiomegaly (heart weighed 560 grams) with concentric left ventricular hypertrophy • Coronary arteriosclerosis, multifocal, moderate, with superimposed coronary thrombosis in the left main stem and right coronary artery • Generalized visceral congestion with pulmonary edema • Smoke inhalation • Scant soot in nares • Soot in upper airway • Blood carboxyhemoglobin level was less than 3%, suggesting the victim was not exposed to excessive carbon monoxide levels	PDF
F2002-46	Jan 04, 2001	Firefighter suffers a heart attack and dies while exercising in firehouse—New York	Atherosclerotic cardiovascular disease	Yes	• Atherosclerotic cardiovascular disease • Coronary artery with marked atherosclerotic stenosis (80%) of the proximal left anterior descending artery • Myocardial infarct, remote (at least 3 months old), septum (microscopic), focal fibrosis of septum adjacent to conduction fibers • Myxoid mitral valve • No evidence of a blood clot (embolus)	PDF
F2002-44	Sep 30, 2002	Parapet wall collapse at auto body shop claims life of career captain and injures career lieutenant and emergency medical technician—Indiana	Extensive blunt-force trauma	Yes	None	PDF

NIOSH REP. #	DATE OF INCIDENT	Title	CAUSE OF DEATH	AUTOPSY	AUTOPSY COMMENTS	PDF LINK
F2002-43	Oct 09, 2002	Firefighter dies after collapse at apartment building fire—kentucky	Ventricular tachyarrhythmia	No	Perform an autopsy on all onduty firefighter fatalities	PDF
F2002-42	Jun 13, 2002	Emergency medical technician killed in single-vehicle crash while responding to structure fire—North Carolina	Blunt-force trauma to the head and neck	Not mentioned	None	PDF
F2002-41	Sep 23, 2002	Career firefighter dies in tanker rollover—North Carolina	Fractured neck due to motor-vehicle incident	Not mentioned	None	PDF
F2002-40	Sep 14, 2002	Career firefighter dies after roof collapse following roof ventilation—Iowa	Smoke inhalation, intra-alveolar hemorrhage, and carbon monoxide intoxication	Yes	Carboxyhemoglobin level, 30.3%	PDF
F2002-39	Sep 05, 2002	Junior volunteer firefighter dies in tanker rollover—Tennessee	Massive head and chest trauma	Not mentioned	None	PDF
F2002-38	Jul 01, 2002	Volunteer captain killed, two firefighters and police officer injured when struck by motor vehicle at highway incident—Minnesota	Multiple blunt-force trauma	Not mentioned	None	PDF
F2002-37	Aug 01, 2002	Volunteer firefighter dies during wildland fire suppression—South Dakota	Second- and third-degree burns over 70 percent of the body	Not mentioned	None	PDF
F2002-36	Aug 08, 2002	Volunteer firefighter dies after being run over by brush truck during grass fire attack—Texas	Craniofacial crush injuries due to a fall from moving vehicle with a secondary run-over	Not mentioned	None	PDF
F2002-35	Jun 08, 2002	Offduty career firefighter dies and another offduty career firefighter is injured after being struck by a truck while assisting at a highway traffic incident—Florida	Multiple blunt-force injuries with compound fractures of the skull	Not mentioned	None	PDF

NIOSH REP. #	DATE OF INCIDENT	Title	CAUSE OF DEATH	AUTOPSY	AUTOPSY COMMENTS	PDF LINK
F2002-34	Jul 30, 2002	Career lieutenant and firefighter die in a flashover during a live-fire training evolution—Florida	Smoke inhalation and thermal injuries	Not mentioned	None	♢PDF
F2002-33	Jan 31, 2002	Firefighter dies during night at fire station—North Carolina	Ischemic heart disease	Yes	• Severe three-vessel atherosclerotic disease • Microscopic examination of one section of the heart reveals multiple small areas of fibrosis. But there is otherwise no evidence of acute or old infarction • A normal sized heart	♢PDF
F2002-32	Jul 04, 2002	Structural collapse at residential fire claims lives of two volunteer fire chiefs and one career firefighter—New Jersey	Fixed compression	Yes	None	♢PDF
F2002-31	Jul 05, 2002	Volunteer firefighter dies due to inadvertent fireworks discharge—North Dakota	Blunt-force injuries of the chest and left upper extremity due to fireworks discharge	Not mentioned	None	♢PDF
F2002-30	Jan 21, 2002	Firefighter suffers probable heart attack at condominium fire—South Carolina	Arteriosclerotic cardio-vascular disease	Yes	• Ischemic fibrosis with scarring, consistent with an remote, healed heart attack (infarct) • Coronary atherosclerosis • Aortic atherosclerosis with ulcerating plaques • A large heart (500 grams)	♢PDF
F2002-28	Apr 27, 2002	Firefighter dies after leaving fire station—Pennsylvania	Sudden cardiac death	No	Perform autopsies on all on-duty firefighter fatalities	♢PDF
F2002-27	May 16, 2001	Firefighter dies during the night at fire station—Missouri	Asphyxiation due to probable seizure	No	Perform an autopsy on all deceased firefighters	♢PDF

NIOSH REP. #	DATE OF INCIDENT	Title	CAUSE OF DEATH	AUTOPSY	AUTOPSY COMMENTS	PDF LINK
F2002-24	Nov 02, 2001	Firefighter suffers fatal heart attack at fire scene—Wisconsin	Per autopsy: acute coronary thrombosis with ischemic arrhythmia Per death certificate: "cardiac arrhythmia" as the immediate cause of death and "CAD" (coronary artery disease) as the underlying cause	Yes	• Severe three-vessel coronary atherosclerosis • Multiple 90-98% stenosis of LAD [left anterior descending] and left circumflex [coronary arteries] • Acute thrombosis, left circumflex artery • Interstitial and perivascular fibrosis, myocardium • Cardiomegaly due to left ventricular hypertrophy • A blood specimen obtained at autopsy contained "<1%" COHb	PDF
F2002-23	Mar 11, 2001	Firefighter suffers fatal heart attack during training—New Mexico	Per autopsy: "myocardial infarction" with "severe hypoxic brain injury" resulting from the cardiac arrest. Per death certificate: "cardiac arrest" as the immediate cause of death, due to "myocardial infarction," with "coronary artery disease" as the underlying cause	Yes	• Acute myocardial infarction, extensive, left ventricle and intraventricular [sic] septum with cardiac arrest and resuscitation • Coronary artery stenosis, bilateral, 95% left coronary and 90% right coronary • Brain death secondary to cardiac arrest (The last finding was based on the hospital record; the autopsy did not include an examination of the brain.)	PDF
F2002-22	Sep 26, 2001	Firefighter suffers fatal heart attack at fire at his residence—Florida	"Acute thrombosis of coronary artery" due to "arteriosclerotic cardiovascular disease"	Yes	Atherosclerotic cardiovascular disease to include • Cardiomegaly (445 grams) with left and right ventricular hypertrophy • Mild to severe calcific atherosclerosis of the coronary arteries • Acute thrombosis of the left anterior descending coronary artery Carboxyhemoglobin (COHb) concentration (a measure of carbon monoxide exposure) in blood obtained at autopsy was 1.4% (a medically insignificant level). There was no soot in the lungs	PDF

NIOSH REP. #	DATE OF INCIDENT	Title	CAUSE OF DEATH	AUTOPSY	AUTOPSY COMMENTS	PDF LINK
F2002-21	May 04, 2002	Junior firefighter killed while responding to fire alarm on his bicycle—Pennsylvania	Multiple injuries	Not mentioned	None	▲PDF
F2002-20	May 03, 2002	Two career firefighters die in four-alarm fire at two-story brick structure—Missouri	Smoke inhalation	Not mentioned	Note: The carbon monoxide level in the blood was noted to be less that 10% in Victim #1, and 47.9% in Victim #2. Victim #1 had third-degree thermal injury over 40% of his body, and Victim #2 had third-degree thermal injury over 18% of his body surface area.	▲PDF
F2002-19	Apr 10, 2002	Firefighter dies during live fire training—North Carolina	Per death certificate: "ischemic heart disease due to coronary artery disease" Per autopsy: "probable cardiac arrhythmia secondary to ischemic heart disease caused by severe coronary artery atherosclerosis"	Yes	• Carboxyhemoglobin level was measured at less than 5%. • Severe coronary atherosclerosis of the left anterior descending and right coronary arteries • Proximal to mid left anterior descending diffuse lesions causing a pinpoint lumen • 80% stenosis of right coronary alternating to minimal throughout length of lumen • Obesity (height of 67 inches and weight of 195 pounds) (Body Mass Index of 30.5 kilograms per square meter (kg/m2)	▲PDF
F2002-18	Apr 11, 2002	Career fire chief dies after being struck by a fire truck at a motor-vehicle incident—Kansas	Chest, abdomen, and skull trauma	Not mentioned	None	▲PDF
F2002-17	Apr 05, 2002	Firefighter dies during the night at fire station—Kansas	Cardiac arrhythmia due to mitral valve prolapse	Yes	• Carboxyhemoglobin level was less than 10%, suggesting that carbon monoxide poisoning was not responsible for his demise • Urine and blood drug screen was negative for illicit drugs and alcohol • Normal sized heart of 360 grams • Moderate myxomatous degeneration, mitral valve with endocardial friction lesions • Foci of myofiber bundle disorder and myocyte disarray of the left ventricle on microscopic examination • Normal coronary arteries • Possible acutely dilated right ventricle	▲PDF

NIOSH REP. #	DATE OF INCIDENT	Title	CAUSE OF DEATH	AUTOPSY	AUTOPSY COMMENTS	PDF LINK
F2002-16	Apr 07, 2002	Volunteer firefighter dies and two are injured in engine rollover—Alabama	Mechanical asphyxia due to blunt head trauma	Not mentioned	None	PDF
F2002-15	Jun 14, 2002	Career firefighter drowns during final dive of training course—Indiana	Drowning	Yes	None	PDF
F2002-14	Jan 10, 2002	Civilian jumps from fourth-story window of burning apartment building and strikes career firefighter—Michigan	County medical examiner listed the cause of death as natural due to a ruptured berry aneurysm; independent review concluded that the firefighter's death was work-related	Yes	The independent review concluded that the firefighter's death was work-related. The report described the cause of death as "…job related, caused by stress and exertion in the performance of his duties. The timeliness of the events on January 10, 2002, and subsequent developments are unquestionably the direct result of the victim's blood pressure which brought on leakage from a pre-existing aneurysm, (sentinel bleed) and ultimately the rupture of the aneurysm, profuse subarachnoid hemorrhage, brain swelling, coma, and death."	PDF
F2002-13	Mar 20, 2002	Volunteer firefighter dies after being struck by motor vehicle on interstate highway—Mississippi	Craniocerebral trauma	Not mentioned	None	PDF
F2002-12	Mar 01, 2002	Volunteer firefighter killed and career chief injured during residential house fire—Tennessee	Asphyxiation	Not mentioned	Carboxyhemoglobin level was listed at 31.8%	PDF
F2002-11	Mar 04, 2002	One career firefighter dies and a captain is hospitalized after floor collapses in residential fire—North Carolina	Multiorgan failure; 80% total body surface burns	Not mentioned	None	PDF
F2002-10	Mar 02, 2002	Volunteer firefighter dies after tanker truck is struck by freight train—kentucky	Multiple blunt-force injuries	Not mentioned	None	PDF

NIOSH REP. #	DATE OF INCIDENT	Title	CAUSE OF DEATH	AUTOPSY	AUTOPSY COMMENTS	PDF LINK
F2002-09	Aug 29, 2001	Firefighter dies while exercising—Florida	Arteriosclerotic and hypertensive heart disease	Yes	• Cardiomegaly (an enlarged heart weighing 700 grams) • Biventricular dilatation • Left ventricular hypertrophy with patchy fibrosis • Scarring consistent with a remote (old) heart attack (myocardial infarction) • Moderate to severe occlusive coronary artery disease (CAD) • Moderate obesity (Body Mass Index of 43 kilograms per square meter [kg/m2]	PDF
F2002-08	Aug 09, 2000	Firefighter dies at kitchen fire—North Carolina	Hypoxia due to pulmonary edema due to cardiomyopathy	Yes	• Dilated cardiomyopathy • Status postmitral valve replacement • Severe pulmonary edema • History of rheumatic heart disease • Cardiomegaly (an enlarged heart weighing 475 grams) • Evidence of chronic ischemia • Morbid obesity (Body Mass Index of 43 kg/m2) • Mild occlusive coronary artery disease (CAD)	PDF
F2002-07	Feb 11, 2002	One career firefighter dies and another is injured after partial structural collapse—Texas	Multiple blunt-force injuries	Not mentioned	None	PDF
F2002-06	Mar 07, 2002	First-floor collapse during residential basement fire claims the lives of two firefighters (career and volunteer) and injures a career firefighter captain—New York	Asphyxia due to the inhalation of smoke and soot	Not mentioned	None	PDF

NIOSH REP. #	DATE OF INCIDENT	Title	CAUSE OF DEATH	AUTOPSY	AUTOPSY COMMENTS	PDF LINK
F2002-05	Dec 14, 2001	Firefighter suffers sudden cardiac death and crashes tanker while responding to a chimney fire—Colorado	Heart failure as the immediate cause of death due to arteriosclerotic heart disease	Yes	• Cardiomegaly • Old healed myocardial infarct [heart attack] involving the interventricular septum and the anterior wall left ventricle • Atherosclerotic coronary artery disease, severe poststent replacement coronary arteries • Atherosclerotic degeneration abdominal portion of the aorta • Small old infarct right cerebral hemisphere [stroke], internal capsule, brain • No evidence for any of the following: epidural, subdural, or subarachnoid [skull] hemorrhage, recent stroke, pulmonary embolus • Blood screen for illicit drug use or alcohol was negative	⅄PDF
F2002-04	Jan 21, 2002	Motor-vehicle incident claims life of volunteer firefighter—Ohio	Laceration of the brain secondary to blunt impacts to the head	Not mentioned	None	⅄PDF
F2002-03	Dec 24, 2001	Firefighter suffers cardiac arrest while responding to a structure fire—Texas	Arteriosclerotic cardiovascular disease	Yes	• Atherosclerosis of the coronary arteries • Cardiomegaly (enlarged heart weighing 600 grams) • Old posterolateral myocardial infarction (prior heart attack) • Bilateral ventricular dilatation • Drug screen was negative for alcohol, illicit drug use, and carbon monoxide	⅄PDF
F2002-02	Dec 15, 2001	Firefighter dies during fire department standby—Arizona	"Probable acute myocardial ischemia" due to "coronary atherosclerosis" due to "diabetes mellitus" as the immediate cause of death and a "history of previous myocardial infarction" as an other significant contributing condition	No	Perform an autopsy on all onduty firefighter fatalities	⅄PDF

NIOSH REP. #	DATE OF INCIDENT	Title	CAUSE OF DEATH	AUTOPSY	AUTOPSY COMMENTS	PDF LINK
F2001-40	Mar 06, 2000	Firefighter suffers cardiac arrest at structure fire—Illinois	Cardiac arrhythmia	Yes	• Atherosclerosis of the coronary arteries • 50% to 75% narrowing of the proximal left anterior descending coronary artery • Cardiomegaly due to left ventricular hypertrophy • Ischemic heart disease	PDF
F2001-39	Nov 19, 2001	Volunteer firefighter killed and an assistant chief injured in tanker truck crash—west Virginia	Blunt-force traumatic head injury	Not mentioned	None	PDF
F2001-38	Sep 25, 2001	Volunteer firefighter dies and two others are injured during live-burn training—New York	Asphyxia due to smoke inhalation	Not mentioned	None	PDF
F2001-37	Jun 18, 2001	Firefighter suffers heart arrhythmia and dies at wildland fire—Washington	Probable cardiac dysrhythmia due to fibrosis of the conduction system of the heart	Yes	• Fibrosis of the conduction system of the heart • Focal moderate to severe atherosclerosis of the proximate left anterior descending (LAD) coronary artery • Mild perivascular fibrosis with adjacent areas of myocyte hypertrophy of the left ventricle and interventricular septum • Possible dilated right ventricle	PDF
F2001-36	Aug 19, 2001	Volunteer firefighter dies when tanker crashes into boulder and tree—Oregon	Blunt-force head trauma	Not mentioned	None	PDF
F2001-35	Oct 13, 2001	Volunteer firefighter drowns during multiagency dive-rescue exercise – Illinois	Drowning	Not mentioned	None	PDF
F2001-34	Aug 12, 2000	Firefighter suffers probable heart attack at fire station—kentucky	Acute myocardial infarction	No	Perform an autopsy on all firefighters who were fatally injured while on duty	PDF
F2001-33	Oct 13, 2001	High-rise apartment fire claims the life of one career firefighter (captain) and injures another career firefighter (captain)—Texas	Asphyxia due to a lack of oxygen.	Not mentioned	The victim's blood level of carboxyhemoglobin measured 18% saturation	PDF

NIOSH REP. #	DATE OF INCIDENT	Title	CAUSE OF DEATH	AUTOPSY	AUTOPSY COMMENTS	PDF LINK
F2001-32	Aug 13, 2001	Firefighter dies at three-alarm structure fire—New York	Hypertrophic and arteriosclerotic cardiac disease	Yes	• A carboxyhemoglobin level (to assess the victim's carbon monoxide exposure) was not performed. • "To assist the investigation of fire-related deaths, we [NIOSH] recommend performing carboxyhemoglobin levels to rule out carbon monoxide poisoning." • Borderline cardiomegaly (an enlarged heart weighing 400 grams) • Dilated cardiac chambers • Pulmonary congestion and edema • Fatty liver • Hepatosplenomegaly	⟫PDF
F2001-31	Apr 23, 2001	Firefighter suffers a fatal heart attack during a training exercise—Michigan	"Acute myocardial infarction" as the immediate cause of death and "arteriosclerotic cardiovascular disease" as the underlying cause	Yes	• Acute myocardial infarction due to arteriosclerotic cardiovascular disease with acute subepicardial antero-lateral myocardial infarction • Coronary atherocalcinosis with maximal manifestation and 75% luminal stenosis in left mid descending coronary artery • Cardiomegaly with dilated cardiomegaly and left ventricular hypertrophy	⟫PDF
F2001-30	Apr 02, 2001	Firefighter suffers fatal heart attack after returning home from fire—Iowa	"Acute myocardial infarction" as the immediate cause of death and "severe arteriosclerotic coronary vascular disease—right coronary artery" as the underlying cause	Yes	• Severe atherosclerotic coronary artery disease—distal right coronary artery • Remote myocardial infarctions. . . . One probably occurred months ago, and another is at least 7 days in age • Respiratory congestion and edema, severe • The result of a blood test for carboxyhemoglobin (an indicator of carbon monoxide exposure) was reported as "0.0%," but the blood was drawn almost 2 hours after Engine 1 departed the fire scene and after 50 minutes of oxygen (which accelerates the clearing of carboxyhemoglobin) administered via endotracheal tube. (At autopsy, the result of another carboxyhemoglobin test was also reported as "0.0%.")	⟫PDF

NIOSH REP. #	DATE OF INCIDENT	Title	CAUSE OF DEATH	AUTOPSY	AUTOPSY COMMENTS	PDF LINK
F2001-29	Feb 12, 2001	Firefighter dies of complications of heart failure suffered at fire scene—Wisconsin	Hypoxia/Ischemia, Acute [of the brain and spinal cord]" as the immediate cause of death, various autopsy findings as intermediate causes, and "Overexertion from responding to a fire call with Fire Dept" as the underlying cause	Yes	• Congestive heart failure • Arteriosclerotic coronary artery disease with luminal occlusion of over 95% in the proximal circumflex which arises aberrantly from the right sinus of Valsalva [and] up to 50% occlusion in anterior interventricular artery, distal circumflex, and posterior intraventricular artery, heart • Anomalous coronary artery distribution • Bronchopneumonia, focal, mild, acute, lungs • Hypoxia/ischemia, acute, varying severity, semiglobal, (cerebral, cerebellar, and spinal cord) • Infarct, acute/recent, middle and inferior temporal gyrus and occipital lobes, bilateral) • Atherosclerosis, moderate to severe, right vertebral artery, basilar artery, temporal branch of right middle cerebral artery • Status postcardiac arrest (pulseless time 6-8 minutes)	⤓PDF
F2001-28	Dec 23, 1995	Firefighter receives severe electrical shock causing cardiac complications, forcing his retirement, and eventually causing his death—Massachusetts	"Arteriosclerotic coronary heart disease" as the immediate cause of death, with "hypertensive heart disease" as a contributing, but not an underlying, cause of death	Yes	• An enlarged heart (cardiomegaly) of 500 grams • All four chambers of the heart were dilated • Thickened left ventricle wall of 1.8 centimeters (normal <1.2 centimeters) • Severe arteriosclerosis with diffuse calcification • Contusion and laceration on the left side of the head without internal injury	⤓PDF
F2001-27	Jun 16, 2001	Career firefighter dies after single-family-residence house fire—South Carolina	Complications due to second- and third-degree burns	Not mentioned	None	⤓PDF
F2001-26	Jul 26, 2001	Career firefighter dies from injuries when stationary fill tank becomes over-pressurized and suffers catastrophic failure—California	Head injury due to blunt impact	Not mentioned	None	⤓PDF

NIOSH REP. #	DATE OF INCIDENT	Title	CAUSE OF DEATH	AUTOPSY	AUTOPSY COMMENTS	PDF LINK
F2001-25	Jun 15, 2001	Firefighter dies after completing job task evaluation—Alabama	Per death certificate: "cardiac arrest" as the immediate cause of death. Per autopsy: "cardiac arrhythmia" due to "myocardial ischemia" due to "coronary artery disease"	Yes	• Moderate coronary atherosclerosis with 50% narrowing of the right coronary artery and the left main artery; 75% narrowing of the left circumflex; 80% narrowing of the left anterior descending coronary artery; left anterior descending coronary artery thrombosis • Acute myocardial ischemia • Severe concentric left ventricular hypertrophy	PDF
F2001-24	Feb 05, 2001	Firefighter suffers sudden cardiac death while exercising during his shift—California	Cardiac dysrhythmia associated with exertion due to dilated cardiomyopathy	Yes	• The heart had a rounded "globoid" configuration rather than the normal pyramidal shape • The heart was enlarged (weighing 480 grams) • All four chambers of the heart were moderately dilated (enlarged) • Both heart ventricles were hypertrophied (thick). • No evidence indicated old or recent myocardial infarctions (heart attacks) • No atherosclerotic changes were evident in the coronary arteries (no blockage in the coronary arteries) • Microscopic examination of the left ventricle, the right ventricle, and the interventricular tissue showed myocyte hypertrophy (muscle cell thickening) characterized by enlarged "box car" nuclei and some patchy interstitial fibrosis (scar tissue between some of the muscle cells) • There was no evidence of vasculitis, myocarditis, sarcoidosis, Fabry's disease, amyloidosis, or hemochromatosis • No illicit drugs, alcohol, or steroids were found • Blood lead level, the urine arsenic level, and the urine mercury level were below the laboratory's level of detection (<3 micrograms (mcg) per deciliter, <15 mcg per liter (mcg/l), <5 mcg/l, respectively). The urine lead level was elevated at 76 mcg/l and when corrected for urine creatinine was 165 mcg per gram creatinine (reference range <50)	PDF

NIOSH REP. #	DATE OF INCIDENT	Title	CAUSE OF DEATH	AUTOPSY	AUTOPSY COMMENTS	PDF LINK
F2001-23	Jun 17, 2001	Hardware store explosion claims the lives of three career firefighters—New York	Victims #1 & #3: massive blunt force trauma Victim #2: Asphyxia due to smoke inhalation.	Not mentioned	None	PDF
F2001-22	Mar 10, 2001	Firefighter dies while on duty—Texas	Hypertensive and arteriosclerotic cardiovascular disease	Yes	• Severe occlusive coronary artery disease with 75% narrowing of the right coronary artery and the diagonal branch; 50% narrowing of the left anterior descending coronary artery and the left circumflex • Cardiomegaly (enlarged heart) • Left ventricular hypertrophy • Pulmonary congestion and edema	PDF
F2001-21	Jan 18, 2001	Firefighter dies in sleep during his work shift—Michigan	Arteriosclerotic cardiovascular disease	Yes	• Significant occlusive coronary artery disease with 85% narrowing of the left main coronary artery, 60% narrowing of the left anterior descending coronary artery, 30% narrowing of the left circumflex, and 25% narrowing of the right coronary artery • Enlarged heart • Left ventricular hypertrophy • Pulmonary congestion and edema	PDF
F2001-20	Dec 02, 1998	Firefighter has sudden cardiac death during training—Texas	Atherosclerotic cardiovascular disease	Yes	• Carboxyhemoglobin level was 1%, suggesting the firefighter was not exposed to excessive concentrations of carbon monoxide (CO). • Atherosclerotic cardiovascular disease • Moderate to severe narrowing of the coronary arteries • Mild to moderate narrowing of the aorta • Mild narrowing of the cerebral arteries • Remote myocardial infarction, multiple, left ventricle	PDF

NIOSH REP. #	DATE OF INCIDENT	Title	CAUSE OF DEATH	AUTOPSY	AUTOPSY COMMENTS	PDF LINK
F2001-19	Feb 19, 2000	Firefighter dies after assisting an injured person—Ohio	Severe occlusive coronary artery disease	Yes	• Severe occlusive coronary artery disease with calcification; 70-80% narrowing of the left anterior descending coronary artery and the right coronary artery; Recent thrombus right coronary artery • Left ventricular hypertrophy • Pulmonary congestion and edema • Congestion of viscera	↧PDF
F2001-18	May 09, 2001	Career firefighter dies after becoming trapped by fire in apartment building—New Jersey	Asphyxiation	Not mentioned	None	↧PDF
F2001-17	Mar 06, 2001	Motor-vehicle incident claims the life of a volunteer assistant chief—Alaska	Multiple blunt-force injuries	Not mentioned	None	↧PDF
F2001-16	Mar 08, 2001	Career firefighter dies after falling through the floor fighting a structure fire at a local residence—Ohio	Complications of thermal burns to 60% of total body surface	Not mentioned	None	↧PDF
F2001-15	Mar 18, 2001	Residential fire claims the lives of two volunteer firefighters and seriously injures an assistant chief—Missouri	Asphyxiation due to smoke inhalation	Not mentioned	None	↧PDF
F2001-14; Grant Number R43-OH-004173	Feb 18, 2001	Firefighter dies after returning from mutual-aid fire call—Connecticut	"Cardiac arrest" is listed as the immediate cause of death, and "hyperlipidemia" and "diabetes mellitus" are listed as contributing factors	No	None	↧PDF
F2001-13	Mar 14, 2001	Supermarket fire claims the life of one career firefighter and critically injures another career firefighter—Arizona	Thermal burns and smoke inhalation	Not mentioned	The victim's carboxyhemoglobin level was listed at 61% at the time of death	↧PDF

NIOSH REP. #	DATE OF INCIDENT	Title	CAUSE OF DEATH	AUTOPSY	AUTOPSY COMMENTS	PDF LINK
F2001-11	Nov 16, 2000	Driver/Operator suffers a cardiac arrest during a wildland fire exercise—Georgia	Per death certificate: atherosclerotic cardiovascular disease Per autopsy: "heart rhythm disturbance (arrhythmia)" as the cause of death and atherosclerotic cardiovascular disease as the contributing factor.	Yes	• Atherosclerotic cardiovascular disease with 70% stenosis, left main coronary artery; 70-90% stenosis, multifocal, left anterior descending coronary artery; complete occlusion, circumflex coronary artery • Vascular congestion, all internal organs	PDF
F2001-10; Purchase Order 0000136411	Dec 23, 2000	Firefighter dies at house fire—New Hampshire	Atherosclerotic and hypertensive cardiovascular disease with cardiac arrhythmia following stress and exertion during firefighting	Yes	• A nontoxic carbon monoxide level • Detectable blood ethyl alcohol • Pulmonary emphysema • Atherosclerosis of the coronary arteries. Specifically, the artery supplying the anterior wall of the heart was focally occluded, and two other major arteries were 40-50% occluded – Acute thrombi (blood clots) were noted	PDF
F2001-09	Feb 25, 2001	Volunteer firefighter dies and another firefighter is injured during wall collapse at fire at local business—Wisconsin	Anoxic encephalopathy and chest compression asphyxia	Not mentioned	None	PDF
F2001-08	Feb 17, 2001	Two volunteer firefighters die fighting a basement fire—Illinois	Asphyxiation caused by inhalation of products of combustion	Yes	None	PDF
F2001-07	Jan 09, 2001	A volunteer firefighter died after being struck by a motor vehicle while directing traffic—New York	Multiple injuries consisting of a massive closed head injury, pulmonary contusion, and chest injury	Not mentioned	None	PDF

NIOSH REP. #	DATE OF INCIDENT	Title	CAUSE OF DEATH	AUTOPSY	AUTOPSY COMMENTS	PDF LINK
F2001-06	Jan 12, 2001	Firefighter dies after the tanker truck he was driving strikes a utility pole and overturns while responding to a grass fire—Kentucky	Blunt-force injuries	Not mentioned	None	PDF
F2001-05	Oct 13, 2000	Driver/Operator dies at his station after responding to three emergency incidents—Massachusetts	Cerebral event as the immediate cause of death, due to cerebral hypoperfusion and coronary artery disease	No	Autopsies should be performed on all onduty firefighters whose death may be cardiovascular-related	PDF
F2001-04	Jan 11, 2001	Volunteer firefighter (lieutenant) killed and one firefighter injured during mobile home fire—Pennsylvania	Asphyxiation	Not mentioned	None	PDF
F2001-03	Dec 28, 2000	Roof collapse injures four career firefighters at a church fire—Arkansas	No deaths	NA	NA	PDF
F2001-02	Aug 17, 2000	A firefighter drowns after attempting to rescue a civilian stranded in flood water—Colorado	Apparent drowning	Not mentioned	None	PDF
F2001-01	Nov 16, 2000	Volunteer firefighter dies and junior firefighter is injured after tanker rollover during water shuttle training exercise—Kentucky	Massive internal injuries	Not mentioned	None	PDF
F2000-45	Jun 23, 2000	Firefighter suffers a heart attack after expressing symptoms while on duty—New Jersey	Arteriosclerotic cardiovascular disease	Yes	• Three arteries are calcified with 10% to 50% obstruction multifocally • Nonadherent thrombus in the right coronary artery • Mild atherosclerosis of the mitral and aortic valves • A 1 cm scar in the posterior wall of the left ventricle, inferiorly	PDF
F2000-44	Nov 25, 2000	Residential house fire claims the life of one career firefighter—Florida	Asphyxia due to smoke inhalation and carbon monoxide poisoning	Yes	Positive carbon monoxide reading of 69.5% was recorded	PDF

NIOSH REP. #	DATE OF INCIDENT	Title	CAUSE OF DEATH	AUTOPSY	AUTOPSY COMMENTS	PDF LINK
F2000-43	Oct 29, 2000	A volunteer assistant chief was seriously injured and two volunteer firefighters were injured while fighting a townhouse fire—Delaware	No deaths	NA	NA	⅊PDF
F2000-42	Feb 13, 2000	Lieutenant suffers a cardiac arrest during a structural drill—Kentucky	"Hypertensive atherosclerotic cardiovascular disease" as the cause of death and "diabetes, hyperlipidemia, tobacco use, and hypertension" as contributing factors	Yes	• Hypertensive atherosclerotic cardiovascular disease • Cardiomegaly • High-grade stenosis (80%) of the circumflex and proximal left anterior descending arteries • Total occlusion of the mid-left anterior descending artery • Stent within the right coronary artery • Acute thrombus of the right coronary artery • Circumferential scarring of the left ventricular free wall and septum • Peripheral vascular disease of the abdominal aorta • Nephrosclerosis • Diabetes mellitus • Hyperlipidemia • Documented history of inferior wall infarction in 1997 • "Documented history of tobacco use" • "Documented history of hypertension"	⅊PDF
F2000-41	Sep 27, 2000	Firefighter dies after being run over by ladder truck while attempting to board—Alabama	Multiple blunt-force injuries	Not mentioned	None	⅊PDF

NIOSH REP. #	DATE OF INCIDENT	Title	CAUSE OF DEATH	AUTOPSY	AUTOPSY COMMENTS	PDF LINK
F2000-40	Aug 06, 2000	Firefighter collapses and dies during exercise training at his fire station—Missouri	Arteriosclerotic heart disease	Yes	• No weight taken at the time of his autopsy • An enlarged heart weighing 570 grams • Left and right ventricular dilatation • Scar formation in the left ventricle (superior aspect of the posterior wall) • Arteriosclerotic heart disease • Fresh clot (thrombus) in the right coronary artery 2 cm from its origin 　–80% narrowing of the distal right coronary artery 　–60% narrowing of the circumflex artery 　–80% narrowing of the left anterior descending artery • Illicit drug and alcohol tests were negative • Presence of a prescription heart medication (Papaverine) • Large white layer of lipid (fat) material [suggesting a recent meal or a lipid disorder	⅄PDF
F2000-39	Apr 29, 2000	A lieutenant dies and three firefighters of a career department were injured when the truck they were responding in was struck by another vehicle—Illinois	Multiple injuries sustained due to a motor-vehicle crash	Not mentioned	None	⅄PDF
F2000-38	Aug 13, 2000	Firefighter/scuba diver dies during training evolution—Indiana	[Pulmonary] barotrauma	Yes	• Extensive air emboli (air bubbles in lung, brain, and heart blood vessels) • Subcutaneous emphysema (air under the skin) and alveolar hemorrhage, as well as pulmonary edema with pink foam in the airways • No pneumothorax. • Toxicology reports listed no finding which contributed to his death	⅄PDF

NIOSH REP. #	DATE OF INCIDENT	Title	CAUSE OF DEATH	AUTOPSY	AUTOPSY COMMENTS	PDF LINK
F2000-37	Jul 18, 2000	Arson investigator dies from injuries sustained from a fall during an arson investigation—Illinois	"Multiple organ system failure" due to, or as a consequence of, (b) "peritonitis with severe hypotension, ischemic necrosis of the liver and kidneys" due to, or as a consequence of, (c) "blunt force trauma of the left chest wall with splenic hematomas and a perforated stress ulcer."	Yes - two	"Blunt force trauma of the left chest wall resulted in a massive intrasplenic hematoma and the development of a stress ulcer. The spleen was not ruptured. The stress ulcer perforated, releasing bowel contents into the peritoneal cavity, which resulted in fibrinous peritonitis. Hypotensive episodes secondary to the peritonitis led to multiple organ system failure, specifically generalized ischemic necrosis of the liver (superimposed upon hepatitis C-induced micronodular cirrhosis), acute tubular necrosis of the kidneys and infarction of the rectosigmoid colon. The combination of liver failure and renal failure constitutes so-called 'multiple organ system failure' and was the immediate cause of death. The underlying cause of death was blunt-force trauma of the left chest wall." Two days later, a second autopsy was performed by a forensic pathologist hired by the victim's widow. This autopsy essentially confirmed the findings of the first autopsy with two additional points mentioned. A left-arm contusion (bruise), a left-trunk contusion, and a left-10th-rib fracture were noted, suggesting that the initial fall on July 18 was quite severe. A second point in this autopsy was that postinjury medications, in addition to chronic hepatitis C virus (HCV) infection and cirrhosis due to HCV, contributed to his liver failure	PDF
F2000-36	Jul 23, 1998	Captain suffers a heart attack at a structure fire and dies 12 days later—Illinois	Myocardial infarction (otherwise known as a heart attack) as the immediate cause of death, due to severe atherosclerotic cardio-vascular disease	Yes	• A large heart (cardiomegaly) weighing 535 grams • An old posterior MI • A recent anterior septal MI • Severe coronary artery disease (CAD) • Stent in the LAD with 60-75% focal occlusion proximally • 70-80% occlusion proximally of the circumflex artery • 50-60% occlusion of the right coronary artery	PDF
F2000-35	Jul 02, 2000	Junior firefighter killed responding to call in his privately owned vehicle (pov)—Pennsylvania	Blunt-force trauma to the head and trunk	Not mentioned	None	PDF

NIOSH REP. #	DATE OF INCIDENT	Title	CAUSE OF DEATH	AUTOPSY	AUTOPSY COMMENTS	PDF LINK
F2000-34	Apr 26, 2000	Lieutenant dies at a fire in a one-and-one-half story dwelling—West Virginia	Atherosclerotic coronary disease	No	A visual inspection, rather than an autopsy, was also completed by the Assistant Medical Examiner. Based on this inspection, the following "pathologic diagnosis" was listed: • "Myocardial infarct (based on circumstances, EKG changes, and serologic testing)" • "Mild chronic obstructive pulmonary disease and hepatitis per history."	⤷PDF
F2000-33	May 27, 2000	Pumper truck rollover claims the life of a volunteer firefighter—Missouri	Cerebral laceration, due to an open skull fracture	Not mentioned	None	⤷PDF
F2000-32	Apr 29, 1998	Driver/Operator/Firefighter suffers a heart attack and dies while supporting fire suppression activities—New York	Hypertensive and atherosclerotic cardiovascular disease	Yes	• Marked coronary artery atherosclerosis – Near occlusive mid-right CAD – Near occlusive proximal left anterior descending CAD • Small circumscribed area (0.5 by 1.2 by 1.0 centimeter) of subepicardial fibrosis in the mid-posterior left ventricular wall (nontransmural). This finding is consistent with a remote (at least 3 months) heart attack in the distribution of the right coronary artery lesion mentioned previously • Left ventricular hypertrophy • No evidence of a blood clot (embolus) in the pulmonary arteries; • His blood carboxyhemoglobin level was less than 3%, suggesting the victim was not exposed to excessive carbon monoxide levels	⤷PDF
F2000-31	Nov 16, 1999	Firefighter collapses at the fire house and subsequently dies due to heart arrhythmia secondary to myocardial sarcoidosis—New Jersey	Granulomatous myocarditis due to sarcoidosis, generalized	Yes	• Cardiac hypertrophy and cardiomegaly (a large heart) • No significant atherosclerotic disease of the coronary arteries • Sarcoidosis (noncaseating granulomas) of the heart, lung, lymph nodes, liver, and spleen	⤷PDF

NIOSH REP. #	DATE OF INCIDENT	Title	CAUSE OF DEATH	AUTOPSY	AUTOPSY COMMENTS	PDF LINK
F2000-30	Sep 28, 1998	District chief dies of a stroke after serving as the Incident Commander at a structure fire—Tennessee	Intracerebral hemorrhage due to hypertension, essential (otherwise known as a "stroke")	No	Perform an autopsy on all onduty firefighter fatalities	⅄PDF
F2000-29	Jan 12, 2000	Firefighter dies as a result of a cardiac arrest at the scene of a structure fire—Maine	Cardiac arrest	Yes	An autopsy was performed; however, medical records were not available to NIOSH personnel at the time of this report	⅄PDF
F2000-28	Aug 20, 1998	Firefighter dies on duty—Tennessee	"acute myocardial infarction due to coronary artery disease" as the immediate cause of death and "diabetes mellitus Type II and renal failure" as other significant conditions	No	Perform an autopsy on all onduty firefighter fatalities	⅄PDF
F2000-27	Apr 30, 2000	Volunteer assistant chief dies during a controlled-burn training evolution—Delaware	Asphyxiation and thermal burns	Not mentioned	Carbon monoxide level was listed at 41%	⅄PDF
F2000-26	Apr 20, 2000	Residential structure fire claims the life of one career firefighter—Alabama	Thermal injuries (full-thickness burns to roughly one-third of body surface area)	Not mentioned	None	⅄PDF
F2000-25	Apr 07, 2000	A volunteer firefighter died and a second was seriously injured after fuel tank explosion—Iowa	Multiple blunt-force injuries to the head, neck, torso, and extremities	Yes	None	⅄PDF
F2000-24	Apr 11, 2000	Firefighter dies during search-and-rescue training—Ohio	Cardiomegaly—acute cardiac arrhythmia	Yes	• Cardiomegaly (enlarged heart) – A weight of 690 grams • Moderate coronary atherosclerosis • 50-60% blockage of the midpoint of his left anterior descending coronary artery • Chronic and acute lung congestion	⅄PDF
F2000-23	Mar 31, 2000	Career firefighter dies and three are injured in a residential garage fire—Utah	Smoke and soot inhalation and acute carbon monoxide intoxication	Not mentioned	• CO levels were at 25% saturation but may not accurately reflect his level due to intubation and resuscitation efforts)	⅄PDF

NIOSH REP. #	DATE OF INCIDENT	Title	CAUSE OF DEATH	AUTOPSY	AUTOPSY COMMENTS	PDF LINK
F2000-22	Mar 15, 2000	Wildland fire claims the life of one volunteer firefighter and injures another firefighter—South Dakota	Sepsis as a result of extensive thermal burns due to a grass fire	Not mentioned	None	PDF

F2000-21	Jun 02, 1998	On-duty driver/operator dies in sleep—Indiana	Probable arrhythmia due to hypertrophic cardiomyopathy, fibrosis of sinoatrial node	Yes	• Since the driver/operator was not engaged in fire suppression activities, his blood was not tested for carbon monoxide poisoning (carboxyhemoglobin levels)
					• Moderate coronary artery disease
					  – Up to 50% stenosis of the proximal circumflex artery
					  – Up to 50% stenosis of the mid left anterior descending coronary artery
					  – Intramyocardial tunneling of coronary artery
					  – Fibrosis of the sinoatrial node
					• Concentric hypertrophic cardiomyopathy
					  – Diminished left ventricular cavity
					  – Dilated right ventricle
					  – Significant myocyte hypertrophy with nuclear enlargement and fiber disarray involving the septum of both ventricles
					• Biventricular and ventricular septum hypertrophy
					• Severely congested lungs bilaterally with edema
					• Hepatosplenomegaly with passive congestion
					• Hashimoto's thyroiditis
						PDF

F2000-20	Feb 10, 1998	Driver/Operator dies at a motor vehicle fire—Wisconsin	"Myocardial infarction" as the immediate cause of death, due to "severe atherosclerotic cardiovascular disease."	Yes	• Severe arteriosclerotic heart disease – High-grade atheromatous plaques of up to 100% occlusion of the left anterior descending and the right coronary arteries • Large remote myocardial infarct in the lateral wall of the left ventricle • Three graft vessels • Cardiomegaly (enlarged heart) (750 gms)	PDF

NIOSH REP. #	DATE OF INCIDENT	Title	CAUSE OF DEATH	AUTOPSY	AUTOPSY COMMENTS	PDF LINK
F2000-19	Mar 17, 2000	Motor-vehicle incident involving amtrak train claims life of career firefighter/engineer—North Carolina	Multiple blunt-force injury from the fire truck/train collision	Not mentioned	None	PDF
F2000-18	Jan 17, 2000	Tanker rollover claims life of volunteer fire chief—Missouri	Massive neck trauma and upper chest trauma	Not mentioned	None	PDF
F2000-17	Feb 11, 2000	A volunteer firefighter/driver was killed and another volunteer firefighter was injured while responding to a motor vehicle incident with injuries—California	Adult respiratory distress syndrome [lung failure]/ischemia and encephalopathy due to motor vehicle crash	Not mentioned	None	PDF
F2000-16	Mar 03, 2000	Arson fire claims the life of one volunteer firefighter and one civilian and severely injures another volunteer firefighter—Michigan	Asphyxiation	Not mentioned	None	PDF
F2000-15	Oct 15, 1997	Battalion chief suffers a heart attack and eventually dies while participating in a fitness program—California	Cardiogenic shock due to enterobacter sepsis due to multiorgan failure due to ischemic cardiomyopathy	No	Perform autopsies on all onduty firefighter fatalities	PDF

F2000-14	Feb 06, 2000	Firefighter dies at a single-family dwelling fire—Iowa	"Occlusive coronary artery disease" as the immediate cause of death due to "atherosclerotic vascular disease"	Yes	• Atherosclerotic vascular disease
					  – Severe coronary artery disease
					  – Right coronary artery, 75% stenotic
					  – Left circumflex artery, 80% stenotic
					  – Left anterior descending artery unavailable due to organ harvesting for donation
					• Focal contraction band necrosis in the right ventricle
					• Acute pulmonary congestion, agonal
					• Pulmonary anthracosis without emphysema
					• Carbon monoxide level of 4%
						PDF

NIOSH REP. #	DATE OF INCIDENT	Title	CAUSE OF DEATH	AUTOPSY	AUTOPSY COMMENTS	PDF LINK
F2000-13	Feb 14, 2000	Restaurant fire claims the life of two career firefighters—Texas	Asphyxia due to smoke inhalation	Yes	None	PDF
F2000-12	Jan 17, 2000	Sector captain suffers fatal heart attack—Texas	"Myocardial infarction" (heart attack) as the immediate cause of death, and "diabetes" as a significant condition	No	None	PDF
F2000-11	Jan 15, 2000	Volunteer firefighter drowns during dry-suit training dive—North Carolina	Severe metabolic acidosis as the result of near drowning	Not mentioned	None	PDF
F2000-10	Oct 28, 1999	A captain and a firefighter die from injuries in a tanker rollover—Indiana	Victim #1: Cardiac arrhythmia Victim #2: Sepsis	Not mentioned	None	PDF
F2000-09	Jan 27, 2000	Volunteer firefighter dies fighting a structure fire at a local residence—Texas	Smoke inhalation	Yes	None	PDF
F2000-07	Jan 17, 2000	Volunteer firefighter dies after 9-foot fall from ladder—Pennsylvania	Blunt-force trauma to the head	Not mentioned	None	PDF
F2000-06	Nov 14, 1999	Tanker rollover results in the death of one volunteer firefighter—Texas	Blunt trauma	Yes	None	PDF

NIOSH REP. #	DATE OF INCIDENT	Title	CAUSE OF DEATH	AUTOPSY	AUTOPSY COMMENTS	PDF LINK

F2000-05	Dec 13, 1999	Firefighter dies at a barn fire—Ohio	"Fatal cardiac arrhythmia" as the immediate cause of death due to "acute thrombotic occlusion of the right coronary artery" (heart attack), "80–90% narrowing of the left anterior descending coronary artery, and severe coronary atherosclerosis."	Yes	• Coronary artery disease
					  – Severe coronary atherosclerosis
					  – Complete acute thrombotic occlusion of the right coronary artery
					  – Narrowing of the left coronary artery
					  – Proximal and middle left anterior descending branch, near complete occlusion
					  – Generalized moderate atherosclerosis
					• Hypertrophy and dilatation of the heart
					• Acute and chronic marked congestion of the lungs
					• Splenomegaly
					The victim's blood was not tested for carbon monoxide poisoning (carboxyhemoglobin levels) although the department requested this be done
						⬇PDF

F2000-04	Dec 22, 1999	Structure fire claims the lives of three career firefighters and three children—Iowa	Victim #1: Smoke inhalation and sudden exposure to intense heat Victim #2: Smoke inhalation and sudden exposure to an extremely hot environment Victim #3: Sudden exposure to intense heat	Not mentioned	Victim #1: 15% Carboxyhemoglobin level Victim #2: 25% Carboxyhemoglobin level Victim #3: 1.0% Carboxyhemoglobin level	⬇PDF
F2000-03	Jul 04, 1999	Captain dies as a result of a cardiac arrest at the scene of a structure fire—Alabama	Myocardial infarction	No	Note: No blood was sent for laboratory analysis during resuscitative efforts. According to post-mortem toxicology forensic specimen analysis, the victim's carboxyhemoglobin level was "negative." An autopsy was not performed Perform an autopsy on all onduty firefighter fatalities	⬇PDF

NIOSH REP. #	DATE OF INCIDENT	Title	CAUSE OF DEATH	AUTOPSY	AUTOPSY COMMENTS	PDF LINK
F2000-02	Jul 06, 1998	Forestry worker dies while bulldozing a fire line at a wildland fire—Alabama	Per death certificate: "cardiac arrest" as the immediate cause of death and "respiratory failure" as the underlying cause Per autopsy: "acute myocardial infarction" (heart attack)	!Yes	• Atherosclerotic cardiovascular disease, trivessel, very severe (Ninety-% occlusion of each of the following arteries: left anterior descending, right, and circumflex arteries) • Myocardial infarctions, old, remote, multiple • Myocardial infarction, recent, acute Toxicological analysis reported a 10-percent carboxyhemoglobin level. In an interview, the head of the Toxicology Section who performed this analysis communicated that the victim's actual carboxyhemoglobin range was within 5 to 10 percent, a range he considered indicative of carbon monoxide exposure during this fire suppression but not indicative of a toxic exposure	ᴧ-PDF
F2000-01	Dec 18, 1999	Motor-vehicle incident claims the life of a volunteer firefighter and injures a volunteer chief—Virginia	Atlanto-occipital dislocation (severe neck spinal cord damage) resulting from motor-vehicle violence	Not mentioned	None	ᴧ-PDF

Appendix C: Examination of Personal Protective Equipment

In some cases, performance of protective clothing and equipment will be a factor in the incident outcome. The condition of all protective clothing and equipment must be properly documented as part of the investigation and can have an impact on a determination for the cause of death during an autopsy. The impounded equipment should be examined at the scene if possible, and again after it has been secured. While at the scene, it is important to note the condition of the equipment in addition to its operational status. Closer examination of equipment may be done following impoundment.

Each item of protective clothing and equipment should be examined carefully. The clothing and equipment always should be photographed. Begin the series of photographs with an overview picture of the item (both front and back). Additional pictures should be taken of every significant component or feature of the clothing and equipment. Pay particular attention to valves, knobs, buttons, and hoses of self-contained breathing apparatus (SCBA) and interface areas for clothing, such as the front closure and collar of pants, ear covers, and visor or goggles of helmets. Table C-1 provides a list of photographs that should be taken for each item of protective clothing and equipment.

Areas of damage or charring also should be photographed. Schematic drawings of protective clothing and equipment should indicate location and extent of the damage. Investigators should take detailed and descriptive notes of all observations. It is as important to note that something was functional or in good condition as it is to note failure or damage. Most performance tests of clothing is destructive. Therefore, it is important to document the condition of the clothing and equipment properly through a detailed written description and photographs/video before any testing is conducted. Note testing always must be authorized by the department before the testing is conducted. Any testing should be identified in advance by:

- the type of test;
- the laboratory where the test will be conducted;
- a description of any samples or specimens removed from the clothing or equipment;
- the purpose of the test; and
- the deposition of any samples that are tested.

Evaluation of SCBA is critical to the investigation. Do not make any adjustments to the SCBA unless absolutely necessary. The positions of straps and knobs should be photographed, and then marked with an indelible marker or grease pencil. Valves should not be opened or shut under any circumstances. If the investigators wish to test breathing air, a SCBA that was filled from the same source at approximately the same time should be used. Several observations should be made at the incident scene:

■ Was the victim wearing SCBA?

■ If not, where was SCBA found in relation to the victim?

■ Was the face piece intact and in place on the victim?

■ Was there pressure remaining in the air cylinder?

■ If so, what were the gauge readings?

■ Did all gauge readings agree?

■ Were valves and regulators in their proper positions?

■ Were support straps and apparatus in their proper configuration, and were they attached as would be expected for normal use?

■ Was there any visible damage to the SCBA (tank, hoses, straps, regulators, mask)?

■ Was there any signs of icing on the regulator (or freeze burns on the face or respiratory tract of the firefighter)?

■ Was any recent maintenance done on the unit?

■ What maintenance records are available on the unit?

■ What were the qualifications of the technician completing the necessary repairs or modifications?

■ Were there any reported problems with this specific unit or with the model?

■ Did the SCBA meet the NFPA 1981 standard in effect at the time of its manufacture?

If investigators have any concerns that the SCBA may have been a factor in the death or injury of firefighters, then the SCBA should be sent to the National Institute for Occupational Safety and Health (NIOSH), in Morgantown, West Virginia. Upon written request, NIOSH investigators will conduct an independent inspection and evaluation of the SCBA. This letter should be sent to NIOSH along with the SCBA to be inspected.

Each member of the fire department who is involved in fire suppression activities is required to have and activate a personal alert safety system (PASS) before entering the hazard area. In some cases, PASS may be integrated with the SCBA and will be activated automatically when the SCBA is worn and used. Investigators should include the following in their observations.

■ Was the victim wearing a PASS device when he or she entered the hazard area?

■ Was it turned on?

■ Is the device capable of being activated with a single gloved hand?

■ Was it functioning when the victim was found?

■ How did the audible alert signal strength compare with a new PASS device with a new battery?

■ Was the victim carrying any other communications equipment (e.g., radio)?

■ Was there any visible damage to the PASS device?

■ Where was it found in relation to the victim?

■ Is it possible that the PASS may have been submerged in water or had the seals to the interior compartment compromised to cause leakage?

■ Was the PASS functional immediately after the incident?

■ When were the PASS batteries last changed?

■ When was the PASS last tested?

■ Was any recent maintenance done on the unit?

■ Did a certified technician complete the necessary repairs or modifications?

■ Were there any previously reported problems with this specific unit or with the model?

■ Did the PASS device meet the NFPA 1982 Standard in effect at the time of its manufacture?

Before beginning the inspection of protective clothing it is important to note the presence and position of the clothing. The following questions, after such documentation, should be addressed:

■ What items of protective clothing was the firefighter wearing (e.g., turnout coat, turnout pants, helmet, gloves, boots, hood, goggles/face shield)?

■ Were all items of protective clothing donned properly?

■ Had the protective clothing been removed?

  – Purposely, by firefighter?

  – Accidently (knocked off)?

  – If so, where was the garment found in relation to the firefighter's body?

■ Were any items of protective clothing removed during rescue efforts?

■ Were any rips, cuts, or tears made during rescue efforts?

■ Did protective clothing meet the appropriate NFPA standard at the time it was manufactured?

■ Every item of protective clothing should be inspected for the following types of wear or damage:

  – Cleanliness, or lack thereof, indicates smoke or chemical exposure.

  – Char, heat damage, and burned areas indicate exposure to excessive heat and/or flame. Areas of damage may cause loss of fabric strength or protective properties. It is particularly important to check all layers of the protective garment.

  – The garment also should be checked for worn or abraded areas, rips, tears, cuts, and fraying. All seams should be checked for broken or missing stitching indicative of seam failure. Signs of discoloration or dye loss also may indicate heat or chemical exposure. Reflective trim should be inspected with a flashlight to ensure that it has not lost its reflective properties.

■ If injuries have occurred, it is especially important to match injury areas on the victim with the areas of the clothing under which the injuries occurred.

Important areas for examination include the following:

■ any gaps in the liner system under the outer shell;

■ types of reinforcements over any injury area; and

■ overall integrity of clothing when worn.

It is important also to ascertain the configuration of clothing as worn (e.g., fastening of closures, position of collar, ear covers, etc.) and whether parts of the ensemble were wet when worn (including wet from an earlier response). It is useful to compare how the clothing was worn by the deceased firefighter during the fatal event, if possible. The sizing of protective clothing and equipment should also be noted in the investigation of personal protective equipment (PPE) performance. Investigation of possible clothing failures must account for differences in clothing performance that occur through wear. It may be useful to compare clothing performance with new or unused items.

Investigation of personal protective clothing and equipment should include the following assessments (as listed by each item):

Protective Helmets

■ helmet outer shell:
- bubbling of shell material,
- delamination of material or soft spots,
- dents, cracks, nicks, gouges, or flaking, and
- loss of surface gloss;

■ helmet inner shell and impact liner:
- warping,
- wear (excessive or unusual),
- broken or missing components, and
- improper installation/attachment of components;

■ suspension system:
- cracked or missing suspension system components,
- torn head band or size-adjustment slots,
- stripped size adjustment ratchet knob,
- signs of excessive wear;

■ crown straps and ear covers:
- improper installation and fit, and
- signs of wear, damage, and excessive heat;

■ damaged chin straps fasteners, slides, and closures; and

■ faceshield/goggles:
- signs of wear, damage, and excessive heat,
- deformation, scratches obscuring vision, and
- damaged fasteners, straps, and closures.

Protective Hoods

■ hood integrity;

■ signs of shrinkage;

■ loss of elasticity;

■ seam integrity; and

■ signs of wear, damage, and charring.

Protective Garments (Coat and Pants)

■ outer shell
- signs of wear, damage, excessive heat, discoloration, or char on fabric,

- seam integrity,
- reinforcement integrity,
- closure system integrity,
- condition of hardware, and
- damage to pockets; items in pockets and their respective condition;

■ moisture barrier and thermal barrier:
- delamination of seams or seals,
- seam integrity/quilt stitching,
- attachment system to the outer shell, and
- signs of wear, damage, excessive heat, discoloration, or char on fabric? (In many cases, it may be necessary to open the liner to determine the condition of the film or coated side of the moisture barrier.);

■ reflective trim
- signs of wear, damage, excessive heat, melt, discoloration, or char on trim,
- seam/stitching integrity, and
- loss of reflectivity; fluorescence;

■ reinforcements (shoulders, elbows, knees, sleeve ends, pant cuffs):
- signs of wear, damage, excessive heat, melt, discoloration, or char on exterior or interior layers,
- seam/stitching and attachment integrity integrity,
- permanent compression;

■ protective wristlets:
- shrinkage,
- loss of elasticity,
- seam integrity, and
- thumbhole elongation;

■ suspenders:
- melting or other heat damage,
- shrinkage,
- loss of elasticity,
- seam integrity, and
- condition of hardware;

Protective Gloves

■ glove integrity;
■ shrinkage;
■ loss of elasticity/flexibility;
■ seam integrity;

■ liner pullout; and

■ signs of wear, damage, excessive heat, discoloration, or char on leather/fabric (both exterior and interior).

Protective Footwear (Rubber)

■ loss of elasticity;

■ delamination of seam seals;

■ material damage;

■ steel toe or shank damage;

■ sole tread wear; and

■ loss of liquid-tight integrity (waterproofness).

Protective Footwear (Leather)

■ seam integrity;

■ material damage (rips, tears, holes);

■ steel toe or shank damage;

■ sole tread wear;

■ loss of liquid-tight integrity (waterproofness);

■ closure system;

■ uniform integrity;

■ seam integrity;

■ material damage (rips, tears, holes);

■ closure system; and

■ signs of wear, damage, excessive heat, discoloration, or char on fabric.

Protective clothing and equipment experts may be required. On occasion, certain garments and equipment may need to be sent out to testing labs for verification that it meets the current set of applicable standards to which the item was certified; however, judgment must be applied if the specific property being measured would be a contributory factor to the firefighter fatality. Other reasons for testing would be to determine whether the item in question was operating properly and, if not, whether it contributed to the incident.

Once the items have been impounded by the investigation team and their condition documented, outside assistance should be requested. All issues involving SCBA testing should be handled by NIOSH. Other protective clothing and equipment testing may be conducted by appropriately qualified experts or independent testing laboratories. Impounded items should be transferred to the testing laboratory following strict chain-of-custody procedures. The testing laboratory should be asked to compare the item performance at the time of the incident with the performance requirements of the appropriate NFPA standard. The independent expert or laboratory may be asked further to determine the range of heat and temperature conditions to which the item may have been exposed. The testing laboratory's report should be summarized in the body of the investigation report and attached as an appendix to the report.

Manufacturer's technical experts may have useful information and should be invited to examine the item in the presence of investigation team members. The manufacturer's written comments should be requested for inclusion in the report. At no time should a manufacturer's representative be given custody of an impounded item or left alone with impounded items.

Table C-1 Recommended Photographs of Protective Clothing and Equipment

Clothing or Equipment Item	Minimum Recommended Photographs
Self-contained breathing apparatus (SCBA)	• Complete SCBA resting on back plate with cylinder shown in front and facepiece off to side • Complete SCBA show harness side • Base of cylinder showing cylinder valve and position • Closeup of manufacturer product label and certification mark • Closeup of cylinder valve and first-stage regulator • Closeup of pressure gauge • Closeup of second-stage regulator • Front of facepiece • Interior of facepiece • Closeup of facepiece exhalation valve • Integrated PASS (if present) • Any specific areas of damage to valves, harness, backplate, hoses, facepiece straps, or accessory items
Personal alert safety system (PASS)	• Complete PASS (top view) • Complete PASS (bottom view) • Closeup of manufacturer product label and certification mark • Closeup of battery panel • Any specific area of damage, particularly for condition of seals in PASS case (if PASS can be opened, the condition of interior components, if warranted by investigation)
Protective helmet	• Front top of helmet (showing shield and emblem) • Left top of helmet • Right top of helmet • Back top of helmet • Interior of helmet showing ear covers (two photographs may be needed to adequately show both sides) • Closeup of product label and certification mark (if possible) • Any specific area of damage such as shell, edge beading, straps, suspension, and ear covers)

Clothing or Equipment Item	Minimum Recommended Photographs
Protective hood	• Left side of hood (lying flat) • Right side of hood (lying flat) • Closeup of product label and certification mark • Any specific area of damage to material or face opening
Protective garments	• Front of garment with closures secured • Back of garment with closures secured • Interior of garment shell (liner removed and shell turned inside out)—both front and back • Removed liner—moisture barrier side—both front and back • Removed liner, turned inside out—thermal barrier side—both front and back • Closeups of all product labels and certification marks on both shell and liner • Any specific area of damage to shell, liner, hardware, trim, reinforcements, and other items on garments (where damage occurs on one side, attempt to photograph other, especially in case of thermal damage)
Protective gloves	• Back side of gloves • Palm side of gloves • Closeup of product label and certification mark • Any specific areas of damage to glove exterior, lining, gauntlet, or wristlet
Protective footwear	• Left side of footwear (standing upright) • Right side of footwear (standing upright) • Footwear soles • Closeup of product label and certification mark (if possible) • Any specific areas of damage to exterior, lining, sole, hardware, or other features of footwear

www.ingramcontent.com/pod-product-compliance
Lightning Source LLC
Chambersburg PA
CBHW081123170526
45165CB00008B/2525